THEORY OF PLATE TECTONICS

Edited by

Judith Rosales

Delve Publishing

Theory of Plate Tectonics

Edited by Judith Rosales

Edition 2018

Delve Publishing

Published by

Delve Publishing LLC

708 3rd Avenue, 6th Floor

New York

NY 10801

United States of America

www.delvepublishing.com

ISBN: 978-1-68095-666-5 (Hardback)

Printed in United States of America

Library of Congress Control Number: 2016961911

About the Editor

Judith Rosales

Prof. Judith is an Ecologist and Environmental Specialist with more than 30 years of academic and consulting experience. She holds a PhD in geography from the University of Birmingham UK and wrote the PhD and Master Academic Programs of Environmental Sciences for the Universidad de Guayana, Venezuela.

Table of Contents

2 THEORY OF PLATE TECTONICS

3 MID-OCEAN RIDGE HISTORY AND FORMATION PROCESSES

4 SUBDUCTION ZONE

5 PROCESS OF MANTLE CONVECTION

6 MOUNTAIN FORMATION PROCESSES

7 AN OVERVIEW OF CURRENT TECTONIC PLATES

8 GEOLOGY OF VENUS

9 GEOLOGY OF MARS

List of Abbreviations

AABW	Antarctic Bottom Water
AAIW	Antarctic Intermediate Water
ACC	Antarctic Circumpolar Current
AMOC	Atlantic Meridional Overturning Circulation
APWP	Apparent Polar Wander Paths
AU	Astronomical Unit
CEIM	Central-East Iranian Microcontinent
CMB	Core-Mantle Boundary
EPR	East Pacific Rise
GMFR	Geomagnetic Field Reversals
GPS	Global Positioning System
GTS	Geological Time Scale
LUA	Last Universal Ancestor
MAR	Mid-Atlantic Ridge
NADW	North Atlantic Deep Water
OIB	Ocean Island Basalts
SFS	Seafloor Spreading Hypothesis
WWSSN	Worldwide Standardized Seismograph Network

Preface

During the past 40 years, a new model has emerged and revolutionized earth sciences like perhaps no other before. The theory of Plate Tectonics is now well-established and forms the basis of our current understanding of the structure and dynamics of our earth. In particular, plate tectonics explains geologic features on the crust over a wide scale, such as the distribution of land and sea, the formation of mountains, earthquakes, and volcanism. Tectonic plates are composed of the rigid outer portion of the earth, called lithosphere. With a thickness of about 100 km, the lithosphere is composed of an upper layer of crust (~7 km thick under the oceans, and ~50 km thick under the continents) and a lower, denser layer of the earth's upper mantle. The rest of the mantle underlying the plates is sufficiently hot to be mobile, although by far the largest part of it is solid. Despite the high temperature, the effect of pressure inside the mantle usually prevents melting.

The rigid lithosphere plates are driven by convection within the mobile asthenosphere. Hot mantle rises beneath mid-oceanic ridges, and cold, denser mantle descends at oceanic trenches. Lateral motion of the lithosphere plates above these circular convection cells is analogous to rigid blocks riding above a rotating conveyor belt. Subduction zones, or convergent margins, are one of the three types of plate boundaries. The others are divergent and transform margins. At a divergent margin, two plates are spreading apart, as at seafloor-spreading ridges or continental rift zones such as the East Africa Rift. Transform margins mark slip-sliding plates, such as California's San Andreas Fault, where the North America and Pacific plates grind past each other with a mostly horizontal motion. The driving force behind plate tectonics is convection in the mantle. Hot material near the earth's core rises, and colder mantle rock sinks. The convection drive plates tectonics through a combination of pushing and spreading apart at mid-ocean ridges and pulling and sinking downward at subduction zones. Scientists continue to study and debate the mechanisms that move the plates. From the deepest ocean trench to the tallest mountain, plate tectonics explains the features and movement of Earth's surface in the present and the past.

Content Coverage

Chapters One and *Two* hypothesizes on the history of earth and theory of plate tectonics, correspondingly. The history of earth concerns the development of the planet earth from its formation to the present day.

Chapters Three and *Four* present the information about mid-ocean ridge formation and subduction process, consistently. Subduction zones create geologic formations such as mountain ranges, ocean trenches, and island arcs, as well as phenomena like earthquakes and volcanoes.

Chapters Five and *Six* are intended to focus on mantle convection and mountain formation processes, separately. Mantle convection is the slow creeping motion of earth's solid silicate mantle caused by convection currents carrying heat from the interior of the Earth to the surface. However, mountain formation processes are associated with large-scale movements of the earth's crust.

Chapters Seven and *Eight* present an overview of current tectonic plates and geology of Venus, separately. Venus is a planet with striking geology. Of all the other planets in the Solar System, it is the one nearest to earth and most like it in terms of mass, but has no magnetic field or recognizable plate tectonic system.

Chapter Nine is geology of Mars that elaborates the scientific study of the surface, crust, and interior of the planet Mars. It emphasizes the composition, structure, history, and physical processes that shape the planet.

CHAPTER 1

History of Earth

Chapter Contents

The planet Earth formed about 4.6 billion years (4,567,000,000 years) ago, in a fully molten state, from the protoplanetary debris disc orbiting the newly formed Sun. Another, smaller Mars-sized planet, called Theia, is thought to have formed in the same orbit as the Earth around this time. After 28 - 34 million years (4.527 - 4.533 billion years ago), orbital oscillations caused Theia to impact the Earth, ejecting a large amount of crust and forming the Moon. That is the best current theory of the Moon's formation, called the Giant impact hypothesis.

INTRODUCTION

Earth formed around 4.54 billion years ago by accretion from the solar nebula. Volcanic outgassing probably created the primordial atmosphere and then the ocean; but the atmosphere contained almost no oxygen and so would have been toxic to most modern life including humans. Much of the Earth was molten because of frequent collisions with other bodies which led to extreme volcanism. A "giant impact" collision with a planet-sized body is thought to have been responsible for forming the Moon. Over time, the Earth cooled, causing the formation of a solid crust, and allowing liquid water to exist on the surface.

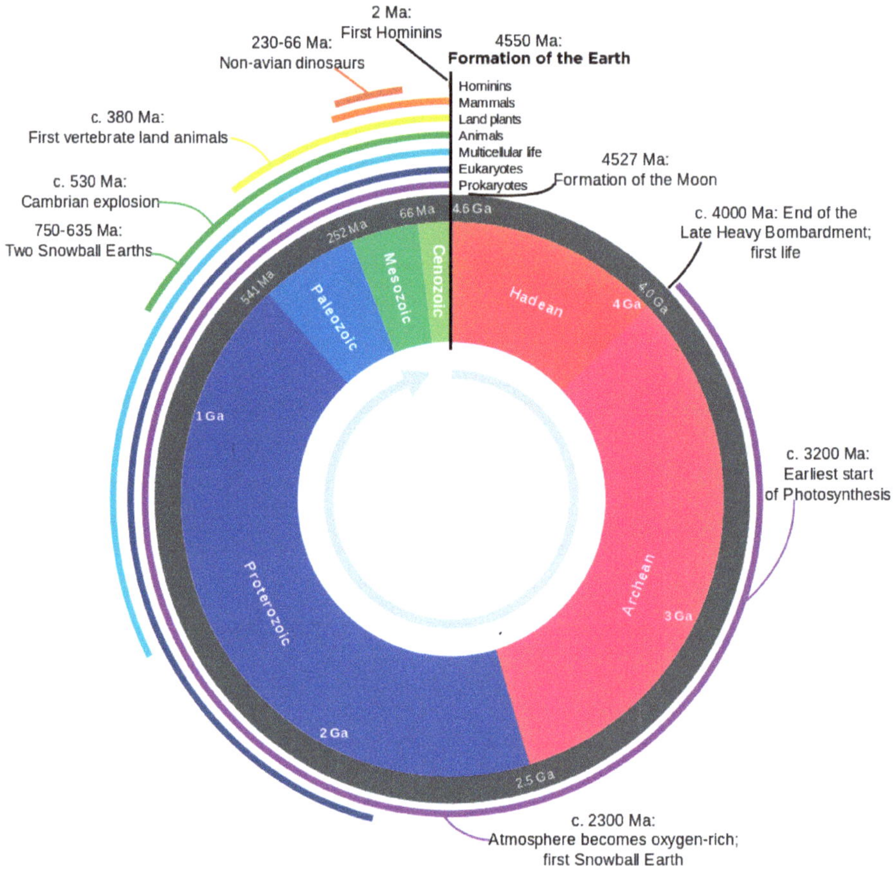

The **history of Earth** concerns the development of the planet Earth from its formation to the present day. Nearly all branches of natural science have contributed to the understanding of the main events of the Earth's past. The age of Earth is approximately one-third of the age of the universe. An immense amount of geological change has occurred in that timespan, accompanied by the emergence of life and its subsequent evolution.

The geological time scale (GTS) clock (see graphic) depicts the larger spans of time from the beginning of the Earth as well as a chronology of some definitive events of Earth history. The Hadean Eon represents time before the reliable (fossil) record of life beginning on Earth; it began with the formation of the planet and ended at 4.0 billion years ago as defined by international convention. The Archean and Proterozoic eons follow; they produced the abiogenesis of life on Earth and then the evolution of early life. The succeeding eon is the Phanerozoic, which is represented by its three component eras: the Palaeozoic; the Mesozoic, which spanned the rise, reign, and climactic extinction of the huge dinosaurs; and the Cenozoic, which presented the subsequent development of dominant mammals on Earth.

Hominins, the earliest direct ancestors of the human clade, rose sometime during the latter part of the Miocene epoch; the precise time marking the first hominins is broadly debated over a current range of 13 to 4 mya. The succeeding Quaternary period is the time of recognizable humans, i.e., the genus *Homo*; but that period's two million-year-plus term of the recent times is too small to be visible at the scale of the GTS graphic. (Notes re the graphic: Ga means "billion years"; Ma, million years.)

The earliest undisputed evidence of life on Earth dates at least from 3.5 billion years ago, during the Eoarchean Era after a geological crust started to solidify following the earlier molten Hadean Eon. There are microbial mat fossils such as stromatolites found in 3.48 billion-year-old sandstone discovered in Western Australia. Other early physical evidence of a biogenic substance is graphite in 3.7 billion-year-old metasedimentary rocks discovered in southwestern Greenland as well as "remains of biotic life found in 4.1 billion-year-old rocks in Western Australia. According to one of the researchers, If life arose relatively quickly on Earth ... then it could be common in the universe.

Living forms derived from photosynthesis appeared between 3.2 and 2.4 billion years ago and began enriching the atmosphere with oxygen. Life remained mostly small and microscopic until about 580 million years ago, when complex multicellular life arose, developed over time, and culminated in the Cambrian Explosion about 541 million years ago. This event drove a rapid diversification of life forms on Earth that produced most of the major phyla known today; and it marked the end of the Proterozoic Eon and the beginning of the Cambrian Period of the Paleozoic Era. More than 99 percent of all species, amounting to over five billion species, that ever lived on Earth are estimated to be extinct. Estimates on the number of Earth's current species range from 10 million to 14 million, of which about 1.2 million have been documented and over 86 percent have not yet been described. More recently, in May 2016, scientists reported that 1 trillion species are estimated to be on Earth currently with only one-thousandth of one percent described.

Geological change has been a constant of Earth's crust since the time of its formation, and biological change since the first appearance of life. Species continue to evolve, taking on new forms, splitting into daughter species or going extinct in the process of adapting or dying in response to ever-changing physical environments. The process of plate tectonics continues to play a dominant role in the shaping of Earth's oceans and continents and the living species they harbor. Changes in the biosphere—now dominated by human activity—continue, in turn, to produce significant effects on the atmosphere and other systems of the Earth's surface, such as the integrity of the ozone layer, the proliferation of greenhouse gases, the conditions of productive soils and clean air and water, and others.

OVERVIEW

In geochronology, time is generally measured in mya (megayears or million years ago), each unit representing the period of approximately 1,000,000 years in the past. The history of Earth is divided into four great eons, starting 4,540 mya with the formation of the planet. Each eon saw the most significant changes in Earth's composition, climate and life. Each eon is subsequently divided into eras, which in turn are divided into periods, which are further divided into epochs.

Eon	Time (mya)	Description
Hadean	4,540–4,000	The Earth is formed out of debris around the solar protoplanetary disk. There is no life. Temperatures are extremely hot, with frequent volcanic activity and hellish environments. The atmosphere is nebular. Possible early oceans or bodies of liquid water. The moon is formed around this time, probably due to a protoplanet's collision into Earth.
Archean	4,000–2,500	Prokaryote life, the first form of life, emerges at the very beginning of this eon, in a process known as abiogenesis. The continents of Ur, Vaalbara and Kenorland may have been formed around this time. The atmosphere is composed of volcanic and greenhouse gases.
Proterozoic	2,500–541	Eukaryotes, a more complex form of life, emerge, including some forms of multicellular organisms. Bacteria began producing oxygen, shaping the third and current of Earth's atmospheres. Plants, later animals and possibly earlier forms of fungi form around this time. The early and late phases of this eon may have undergone "Snowball Earth" periods, in which all of the planet suffered below-zero temperatures. The early continents of Columbia, Rodinia and Pannotia may have formed around this time, in that order.
Phanerozoic	541–present	Complex life, including vertebrates, began to dominate the Earth's ocean in a process known as the Cambrian explosion. Pangaea forms and later dissolves into Laurasia and Gondwana. Gradually, life expanded to land and all familiar forms of plants, animals and fungi began appearing, including annelids, insects and reptiles. Several mass extinctions occur, among which birds, the descendants of dinosaurs, and more recently mammals emerge. The rise and evolution of the human species occur at the very recent phases of this eon.

GEOLOGIC TIME SCALE

The history of the Earth can be organized chronologically according to the geologic time scale, which is split into intervals based on stratigraphic analysis. The following four timelines show the geologic time scale. The first shows the entire time from the formation of the Earth to the present, but this compresses the most recent eon. Therefore, the second scale shows the most recent eon with an expanded scale. The second scale compresses the most recent era, so the most recent era is expanded in the third scale. The third scale compresses the most recent period, so the most recent period is expanded in the fourth scale.

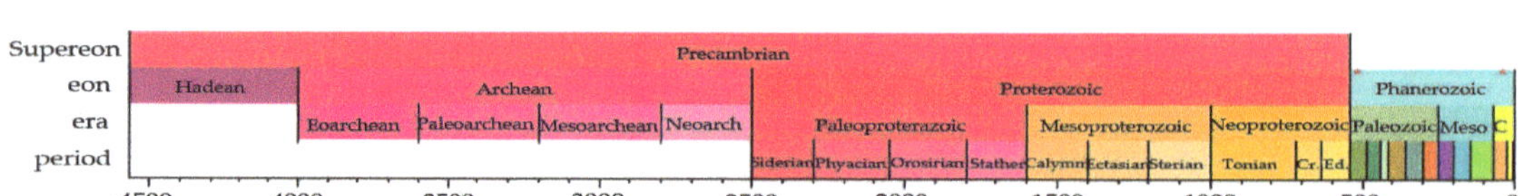

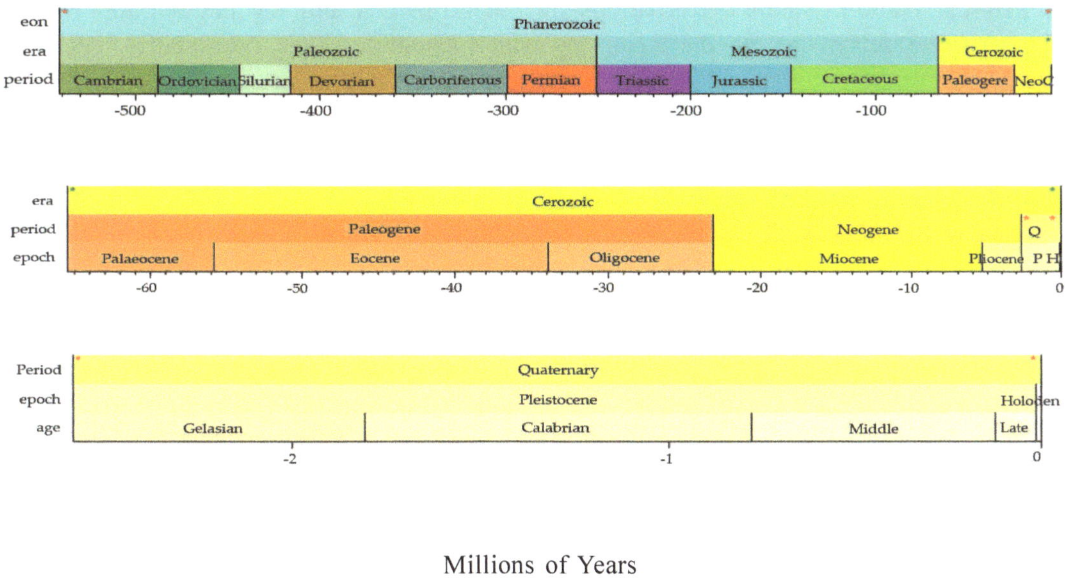

SOLAR SYSTEM FORMATION

An artist's rendering of a protoplanetary disk.

The standard model for the formation of the Solar System (including the Earth) is the solar nebula hypothesis. In this model, the Solar System formed from a large, rotating cloud of interstellar dust and gas called the solar nebula. It was composed of hydrogen and helium created shortly after the Big Bang 13.8 Ga (billion years ago) and heavier elements ejected by supernovae. About 4.5 Ga, the nebula began a contraction that may have been triggered by the shock wave of a nearby supernova. A shock wave would have also made the nebula rotate. As the cloud began to accelerate, its angular momentum, gravity and inertia flattened it into a protoplanetary disk perpendicular to its axis of rotation. Small perturbations due to collisions and the angular momentum of other large debris created the means by which kilometer-sized protoplanets began to form, orbiting the nebular center.

The center of the nebula, not having much angular momentum, collapsed rapidly, the compression heating it until nuclear fusion of hydrogen into helium began. After more contraction, a T Tauri star ignited and evolved into the Sun. Meanwhile, in the outer part of the nebula gravity caused matter to condense around density perturbations and dust particles, and the rest of the protoplanetary disk began separating into rings. In a process known as runaway accretion, successively larger fragments of dust and debris clumped together to form planets. Earth formed in this manner about 4.54 billion years ago (with an uncertainty of 1%) and was largely completed within 10–20 million years. The solar wind of the newly formed T Tauri star cleared out most of the material in the disk that had not already condensed into larger bodies. The same process is expected to produce accretion disks around virtually all newly forming stars in the universe, some of which yield planets.

The proto-Earth grew by accretion until its interior was hot enough to melt the heavy, siderophile metals. Having higher densities than the silicates, these metals sank. This so-called *iron catastrophe* resulted in the separation of a primitive mantle and a (metallic) core only 10 million years after the Earth began to form, producing the layered structure of Earth and setting up the formation of Earth's magnetic field. J. A. Jacobs was the first to suggest that the inner core—a solid center distinct from the liquid outer core—is freezing and growing out of the liquid outer core due to the gradual cooling of Earth's interior (about 100 degrees Celsius per billion years).

HADEAN AND ARCHEAN EONS

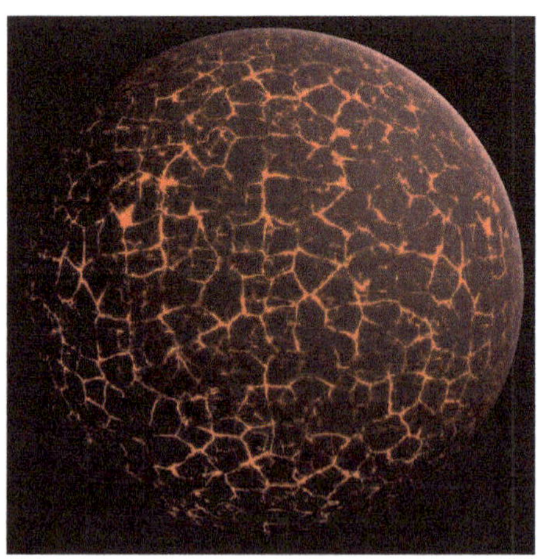

Artist's conception of Hadean Eon Earth, when it was much hotter and inhospitable to all forms of life.

The first eon in Earth's history, the *Hadean*, begins with the Earth's formation and is followed by the *Archean* eon at 3.8 Ga. The oldest rocks found on Earth date to about 4.0 Ga, and the oldest detritalzircon crystals in rocks to about 4.4 Ga, soon after the formation of the Earth's crust and the Earth itself. The giant impact hypothesis for the Moon's formation states that shortly after formation of an initial crust, the proto-Earth was impacted by a smaller protoplanet, which ejected part of the mantle and crust into space and created the Moon. From crater counts on other celestial bodies it is inferred that a period of intense meteorite impacts, called the *Late Heavy Bombardment*, began about 4.1 Ga, and concluded around 3.8 Ga, at the end of the Hadean. In addition, volcanism was severe due to the large heat flow and geothermal gradient.Nevertheless, detrital zircon crystals dated to 4.4 Ga show evidence of having undergone contact with liquid water, suggesting that the Earth already had oceans or seas at that time.

By the beginning of the Archean, the Earth had cooled significantly. Present life forms could not have survived at Earth's surface, because the Archean atmosphere lacked oxygen hence had no ozone layer to block ultraviolet light. Nevertheless, it is believed that primordial life began to evolve by the early Archean, with candidate fossils dated to around 3.5 Ga. Some scientists even speculate that life could have begun during the early Hadean, as far back as 4.4 Ga, surviving the possible Late Heavy Bombardment period in hydrothermal vents below the Earth's surface.

Formation of the Moon

Earth's only natural satellite, the Moon, is larger relative to its planet than any other satellite in the solar system During the Apollo program, rocks from the Moon's surface were brought to Earth. Radiometric dating of these rocks has shown that the Moon is 4.53 ± .01 billion years old, formed at least 30 million years after the solar system. New evidence suggests the Moon formed even later, 4.48 ± 0.02 Ga, or 70–110 million years after the start of the Solar System.

Artist's impression of the enormous collision that probably formed the Moon.

Theories for the formation of the Moon must explain its late formation as well as the following facts. First, the Moon has a low density (3.3 times that of water, compared to 5.5 for the earth) and a small metallic core. Second, there is virtually no water or other volatiles on the moon. Third, the Earth and Moon have the same oxygen isotopic signature (relative abundance of the oxygen isotopes). Of the theories that have been proposed to account for these phenomena, only one is widely accepted: The *giant impact hypothesis* proposes that the Moon originated after a body the size of Mars (sometimes named Theia) struck the proto-Earth a glancing blow.

The collision released about 100 million times more energy than the more recent Chicxulub impact that is believed to have caused the extinction of the dinosaurs. It was enough to vaporize some of the Earth's outer layers and melt both bodies. A portion of the mantle material was ejected into orbit around the Earth. The giant impact hypothesis predicts that the Moon was depleted of metallic material, explaining its abnormal composition. The ejecta in orbit around the Earth could have condensed into a single body within a couple of weeks. Under the influence of its own gravity, the ejected material became a more spherical body: the Moon.

First Continents

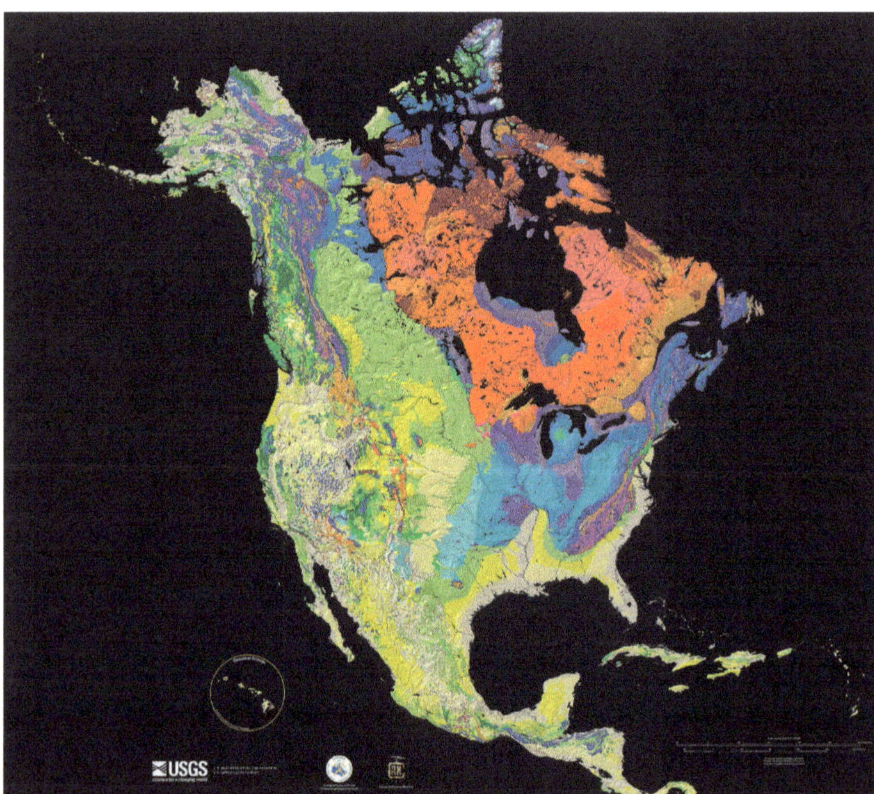

Geologic map of North America, color-coded by age. The reds and pinks indicate rock from the Archean.

Mantle convection, the process that drives plate tectonics today, is a result of heat flow from the Earth's interior to the Earth's surface. It involves the creation of rigid tectonic plates at mid-oceanic ridges. These plates are destroyed by subduction into the mantle at subduction zones. During the early Archean (about 3.0 Ga) the mantle was much hotter than today, probably around 1,600 °C (2,910 °F), so convection in the mantle was faster. Although a process similar to present-day plate tectonics did occur, this would have gone faster too. It is likely that during the Hadean and Archean, subduction zones were more common, and therefore tectonic plates were smaller.

The initial crust, formed when the Earth's surface first solidified, totally disappeared from a combination of this fast Hadean plate tectonics and the intense impacts of the Late Heavy Bombardment. However, it is thought that it was basaltic in composition, like today's oceanic crust, because little crustal differentiation had yet taken place. The first larger pieces of continental crust, which is a product of differentiation of lighter elements during partial melting in the lower crust, appeared at the end of the Hadean, about 4.0 Ga. What is left of these first small continents are called cratons. These pieces of late Hadean and early Archean crust form the cores around which today's continents grew.

The oldest rocks on Earth are found in the North American craton of Canada. They are tonalites from about 4.0 Ga. They show traces of metamorphism by high temperature, but also sedimentary grains that have been rounded by erosion during transport by water, showing that rivers and seas existed then. Cratons consist primarily of two alternating types of terranes. The first are so-called greenstone belts, consisting of low-grade metamorphosed sedimentary rocks. These greenstones are similar to the sediments today found in oceanic trenches, above subduction zones. For this reason, greenstones are sometimes seen as evidence for subduction during the Archean. The second type is a complex of felsicmagmatic rocks. These rocks are mostly tonalite, trondhjemite or granodiorite, types of rock similar in composition to granite (hence such terranes are called TTG-terranes). TTG-complexes are seen as the relicts of the first continental crust, formed by partial melting in basalt.

Oceans and Atmosphere

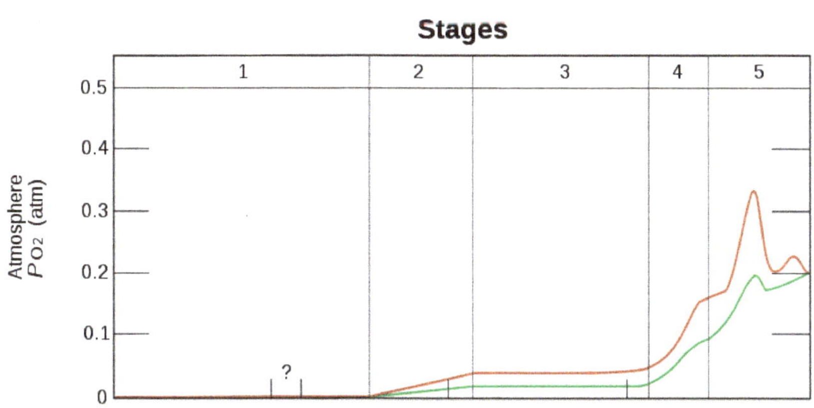

Graph showing range of estimated partial pressure of atmospheric oxygen through geologic time

Earth is often described as having had three atmospheres. The first atmosphere, captured from the solar nebula, was composed of light (atmophile) elements from the solar nebula, mostly hydrogen and helium. A combination of the solar wind and Earth's heat would have driven off this atmosphere, as a result of which the atmosphere is now depleted of these elements compared to cosmic abundances. After the impact which created the moon, the molten Earth released volatile gases; and later more gases were released by volcanoes, completing a second atmosphere rich in greenhouse gases but poor in oxygen. Finally, the third atmosphere, rich in oxygen, emerged when bacteria began to produce oxygen about 2.8 Ga.

In early models for the formation of the atmosphere and ocean, the second atmosphere was formed by outgassing of volatiles from the Earth's interior. Now it is considered likely that many of the volatiles were delivered during accretion by a process known as *impact degassing* in which incoming bodies vaporize on impact. The ocean and atmosphere would therefore have started to form even as the Earth formed. The new atmosphere probably contained water vapor, carbon dioxide, nitrogen, and smaller amounts of other gases.

Planetesimals at a distance of 1 astronomical unit (AU), the distance of the Earth from the Sun, probably did not contribute any water to the Earth because the solar nebula was too hot for ice to form and the hydration of rocks by water vapor would have taken too long. The water must have been supplied by meteorites from the outer asteroid belt and some large planetary embryos from beyond 2.5 AU. Comets may also have contributed. Though most comets are today in orbits farther away from the Sun than Neptune, computer simulations show they were originally far more common in the inner parts of the solar system.

As the Earth cooled, clouds formed. Rain created the oceans. Recent evidence suggests the oceans may have begun forming as early as 4.4 Ga. By the start of the Archean eon they already covered the Earth. This early formation has been difficult to explain because of a problem known as the faint young Sun paradox. Stars are known to get brighter as they age, and at the time of its formation the Sun would have been emitting only 70% of its current power. Thus, the Sun has become 30% brighter in the last 4.5 billion years. Many models predict that the Earth would have been covered in ice. A likely solution is that there was enough carbon dioxide and methane to produce a greenhouse effect. The carbon dioxide would have been produced by volcanoes and the methane by early microbes. Another greenhouse gas, ammonia, would have been ejected by volcanos but quickly destroyed by ultraviolet radiation.

Origin of Life

The first step in the emergence of life may have been chemical reactions that produced many of the simpler organic compounds, including nucleobases and amino acids that are the building blocks of life. An experiment in 1953 by Stanley Miller and Harold Urey showed that such molecules could form in an atmosphere of water, methane, ammonia and hydrogen with the aid of sparks to mimic the effect of lightning. Although the atmospheric composition was probably different from the composition used by Miller and Urey, later experiments with more realistic compositions also managed to synthesize organic molecules. Recent computer simulations have even shown that extraterrestrial organic molecules could have formed in the protoplanetary disk before the formation of the Earth.

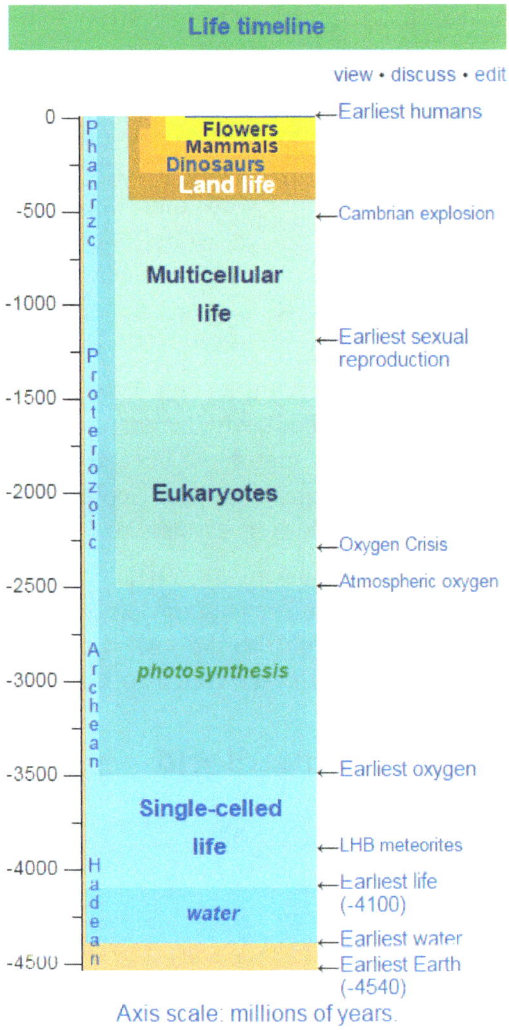

One of the reasons for interest in the early atmosphere and ocean is that they form the conditions under which life first arose. There are many models, but little consensus, on how life emerged from non-living chemicals; chemical systems that have been created in the laboratory still fall well short of the minimum complexity for a living organism.

The next stage of complexity could have been reached from at least three possible starting points: self-replication, an organism's ability to produce offspring that are very similar to itself; metabolism, its ability to feed and repair itself; and external cell membranes, which allow food to enter and waste products to leave, but exclude unwanted substances.

Replication first: RNA world

Even the simplest members of the three modern domains of life use DNA to record their "recipes and a complex array of RNA and protein molecules to "read" these instructions and use them for growth, maintenance and self-replication.

The discovery that a kind of RNA molecule called a ribozyme can catalyze both its own replication and the construction of proteins led to the hypothesis that earlier life-forms were based entirely on RNA. They could have formed an RNA world in which there were individuals but no species, as mutations and horizontal gene transfers would have meant that the offspring in each generation were quite likely to have different genomes from those that their parents started with. RNA would later have been replaced by DNA, which is more stable and therefore can build longer genomes, expanding the range of capabilities a single organism can have. Ribozymes remain as the main components of ribosomes, the protein factories of modern cells.

Although short, self-replicating RNA molecules have been artificially produced in laboratories, doubts have been raised about whether natural non-biological synthesis of RNA is possible. The earliest ribozymes may have been formed of simpler nucleic acids such as PNA, TNA or GNA, which would have been replaced later by RNA. Other pre-RNA replicators have been posited, including crystals and even quantum systems.

In 2003 it was proposed that porous metal sulfide precipitates would assist RNA synthesis at about 100 °C (212 °F) and ocean-bottom pressures near hydrothermal vents. In this hypothesis, lipid membranes would be the last major cell components to appear and until they did the proto-cells would be confined to the pores.

Metabolism first: iron–Sulfur World

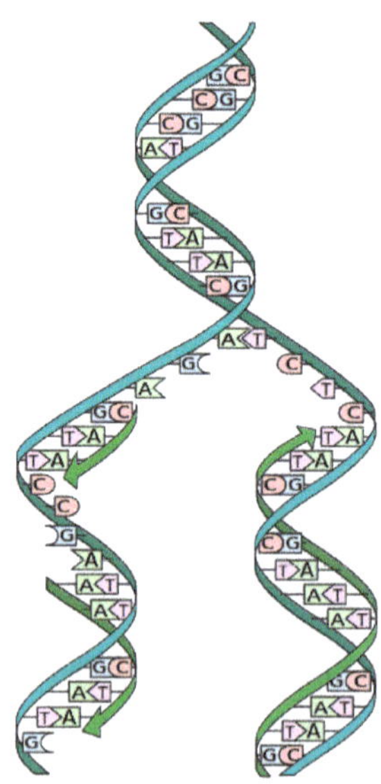

The replicator in virtually all known life is deoxyribonucleic acid. DNA is far more complex than the original replicator and its replication systems are highly elaborate.

Another long-standing hypothesis is that the first life was composed of protein molecules. Amino acids, the building blocks of proteins, are easily synthesized in plausible prebiotic conditions, as are small peptides (polymers of amino acids) that make good catalysts. A series of experiments starting in 1997 showed that amino acids and peptides could form in the presence of carbon monoxide and hydrogen sulfide with iron sulfide and nickel sulfide as catalysts. Most of the steps in their assembly required temperatures of about 100 °C (212 °F) and moderate pressures, although one stage required 250 °C (482 °F) and a pressure equivalent to that found under 7 kilometers (4.3 mi) of rock. Hence, self-sustaining synthesis of proteins could have occurred near hydrothermal vents.

A difficulty with the metabolism-first scenario is finding a way for organisms to evolve. Without the ability to replicate as individuals, aggregates of molecules would have compositional genomes (counts of molecular species in the aggregate) as the target of natural selection. However, a recent model shows that such a system is unable to evolve in response to natural selection.

Membranes first: Lipid world

It has been suggested that double-walled "bubbles" of lipids like those that form the external membranes of cells may have been an essential first step. Experiments that simulated the conditions of the early Earth have reported the formation of lipids, and these can spontaneously form liposomes, double-walled bubbles, and then reproduce themselves. Although they are not intrinsically information-carriers as nucleic acids are, they would be subject to natural selection for longevity and reproduction. Nucleic acids such as RNA might then have formed more easily within the liposomes than they would have outside.

The clay Theory

Some clays, notably montmorillonite, have properties that make them plausible accelerators for the emergence of an RNA world: they grow by self-replication of their crystalline pattern, are subject to an analog of natural selection (as the clay "species" that grows fastest in a particular environment rapidly becomes dominant), and can catalyze the formation of RNA molecules. Although this idea has not become the scientific consensus, it still has active supporters.

Research in 2003 reported that montmorillonite could also accelerate the conversion of fatty acids into "bubbles", and that the bubbles could encapsulate RNA attached to the clay. Bubbles can then grow by absorbing additional lipids and dividing. The formation of the earliest cells may have been aided by similar processes.

A similar hypothesis presents self-replicating iron-rich clays as the progenitors of nucleotides, lipids and amino acids.

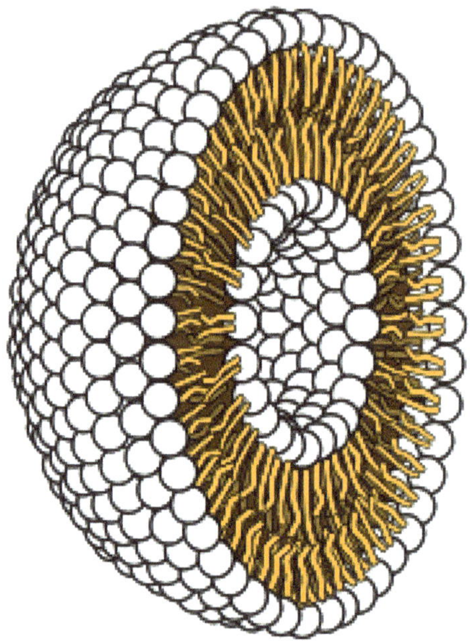

Cross-section through a liposome.

Last Universal Ancestor

It is believed that of this multiplicity of protocells, only one line survived. Current phylogenetic evidence suggests that the last universal ancestor (LUA) lived during the early Archean eon, perhaps 3.5 Ga or earlier. This LUA cell is the ancestor of all life on Earth today. It was probably a prokaryote, possessing a cell membrane and probably ribosomes, but lacking a nucleus or membrane-bound organelles such as mitochondria or chloroplasts. Like all modern cells, it used DNA as its genetic code, RNA for information transfer and protein synthesis, and enzymes to catalyze reactions. Some scientists believe that instead of a single organism being the last universal common ancestor, there were populations of organisms exchanging genes by lateral gene transfer.

PROTEROZOIC EON

The Proterozoic eon lasted from 2.5 Ga to 542 Ma (million years) ago. In this time span, cratons grew into continents with modern sizes. The change to an oxygen-rich atmosphere was a crucial development. Life developed from prokaryotes into eukaryotes and multicellular forms. The Proterozoic saw a couple of severe ice ages called snowball Earths. After the last Snowball Earth about 600 Ma, the evolution of life on Earth accelerated. About 580 Ma, the Ediacara biota formed the prelude for the Cambrian Explosion.

Oxygen Revolution

Lithified stromatolites on the shores of Lake Thetis, Western Australia. Archean stromatolites are the first direct fossil traces of life on Earth.

A banded iron formation from the 3.15 Ga Moories Group, Barberton Greenstone Belt, South Africa. Red layers represent the times when oxygen was available, gray layers were formed in anoxic circumstances.

The earliest cells absorbed energy and food from the surrounding environment. They used fermentation, the breakdown of more complex compounds into less complex compounds with less energy, and used the energy so liberated to grow and reproduce.

Fermentation can only occur in an *anaerobic* (oxygen-free) environment. The evolution of photosynthesis made it possible for cells to manufacture their own food.

Most of the life that covers the surface of the Earth depends directly or indirectly on photosynthesis. The most common form, oxygenic photosynthesis, turns carbon dioxide, water and sunlight into food. It captures the energy of sunlight in energy-rich molecules such as ATP, which then provide the energy to make sugars. To supply the electrons in the circuit, hydrogen is stripped from water, leaving oxygen as a waste product. Some organisms, including purple bacteria and green sulfur bacteria, use an anoxygenic form of photosynthesis that use alternatives to hydrogen stripped from water as electron donors; examples are hydrogen sulfide, sulfur and iron. Such extremophile organisms are restricted to otherwise inhospitable environments such as hot springs and hydrothermal vents.

The simpler anoxygenic form arose about 3.8 Ga, not long after the appearance of life. The timing of oxygenic photosynthesis is more controversial; it had certainly appeared by about 2.4 Ga, but some researchers put it back as far as 3.2 Ga. The latter "probably increased global productivity by at least two or three orders of magnitude". Among the oldest remnants of oxygen-producing lifeforms are fossil stromatolites.

At first, the released oxygen was bound up with limestone, iron, and other minerals. The oxidized iron appears as red layers in geological strata called banded iron formations that formed in abundance during the Siderian period (between 2500 Ma and 2300 Ma). When most of the exposed readily reacting minerals were oxidized, oxygen finally began to accumulate in the atmosphere. Though each cell only produced a minute amount of oxygen, the combined metabolism of many cells over a vast time transformed Earth's atmosphere to its current state. This was Earth's third atmosphere.

Some oxygen was stimulated by incoming ultraviolet radiation to form ozone, which collected in a layer near the upper part of the atmosphere. The ozone layer absorbed, and still absorbs, a significant amount of the ultraviolet radiation that once had passed through the atmosphere. It allowed cells to colonize the surface of the ocean and eventually the land: without the ozone layer, ultraviolet radiation bombarding land and sea would have caused unsustainable levels of mutation in exposed cells.

Photosynthesis had another major impact. Oxygen was toxic; much life on Earth probably died out as its levels rose in what is known as the *oxygen catastrophe*. Resistant forms survived and thrived, and some developed the ability to use oxygen to increase their metabolism and obtain more energy from the same food.

Snowball Earth

The natural evolution of the Sun made it progressively more luminous during the Archean and Proterozoic eons; the Sun's luminosity increases 6% every billion years. As a result, the Earth began to receive more heat from the Sun in the Proterozoic eon. However, the Earth did not get warmer. Instead, the geological record seems to suggest it cooled dramatically during the early Proterozoic. Glacial deposits found in South Africa date back to 2.2 Ga, at which time, based on paleomagnetic evidence, they must have been located near the equator. Thus, this glaciation, known as the Makganyene glaciation, may have been global. Some scientists suggest this was so severe that the Earth was totally frozen over from the poles to the equator, a hypothesis called Snowball Earth. The ice age around 2.3 Ga could have been directly caused by the increased oxygen concentration in the atmosphere, which caused the decrease of methane (CH_4) in the atmosphere. Methane is a strong greenhouse gas, but with oxygen it reacts to form CO_2, a less effective greenhouse gas. When free oxygen became available in the atmosphere, the concentration of methane could have decreased dramatically, enough to counter the effect of the increasing heat flow from the Sun.

However, the term Snowball Earth is more commonly used to describe later extreme ice ages during the Cryogenian period. There were four periods, each lasting about 10 million years, between 750 and 580 million years ago, when the earth is thought to have been covered with ice apart from the highest mountains, and average temperatures were about −50 °C (−58 °F). The snowball may have been partly due to the location of the supercontintent Rodinia straddling the Equator. Carbon dioxide combines with rain to weather rocks to form carbonic acid, which is then washed out to sea, thus extracting the greenhouse gas from the atmosphere. When the continents are near the poles, the advance of ice covers the rocks, slowing the reduction in carbon dioxide, but in the Cryogienian the weathering of Rodinia was able to continue unchecked until the ice advanced to the tropics. The process may have finally been reversed by the emission of carbon dioxide from volcanoes or the destabilization of methane gas hydrates. According to the alternative Slushball Earth theory, even at the height of the ice ages there was still open water at the Equator.

Emergence of Eukaryotes

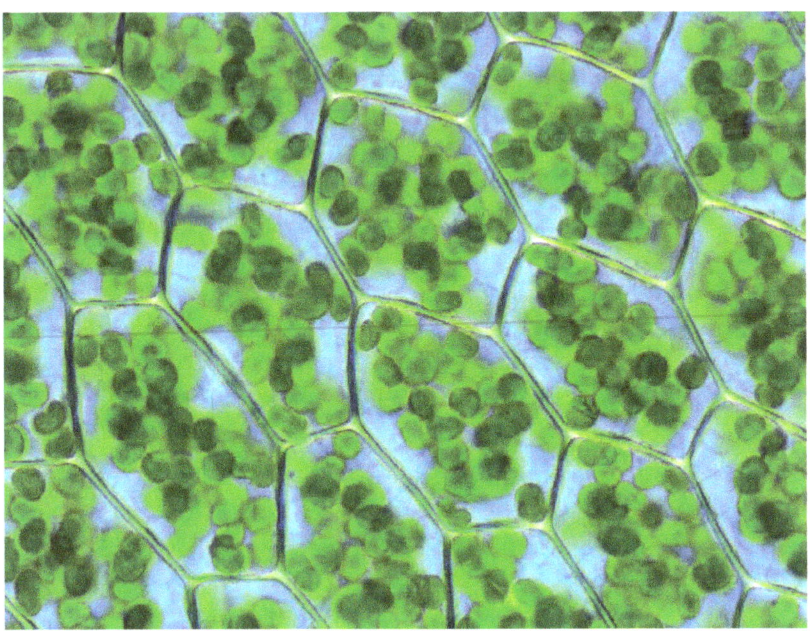

Chloroplasts in the cells of a moss.

Modern taxonomy classifies life into three domains. The time of the origin of these domains is uncertain. The Bacteria domain probably first split off from the other forms of life (sometimes called Neomura), but this supposition is controversial. Soon after this, by 2 Ga, the Neomura split into the Archaea and the Eukarya. Eukaryotic cells (Eukarya) are larger and more complex than prokaryotic cells (Bacteria and Archaea), and the origin of that complexity is only now becoming known.

Around this time, the first proto-mitochondrion was formed. A bacterial cell related to today's *Rickettsia*, which had evolved to metabolize oxygen, entered a larger prokaryotic cell, which lacked that capability. Perhaps the large cell attempted to digest the smaller one but failed (possibly due to the evolution of prey defenses). The smaller cell may have tried to parasitize the larger one. In any case, the smaller cell survived inside the larger cell. Using oxygen, it metabolized the larger cell's waste products and derived more energy. Part of this excess energy was returned to the host. The smaller cell replicated inside the larger one. Soon, a stable symbiosis developed between the large cell and the smaller cells inside it. Over time, the host cell acquired some genes from the smaller cells, and the two kinds became dependent on each other: the larger cell could not survive without the energy produced by the smaller ones, and these in turn could not survive without the raw materials provided by the larger cell. The whole cell is now considered a single organism, and the smaller cells are classified as organelles called mitochondria.

A similar event occurred with photosynthetic cyanobacteria entering large heterotrophic cells and becoming chloroplasts. Probably as a result of these changes, a line of cells capable of photosynthesis split off from the other eukaryotes more than 1 billion years ago. There were probably several such inclusion events. Besides the well-established endosymbiotic theory of the cellular origin of mitochondria and chloroplasts, there are theories that cells led to peroxisomes, spirochetes led to cilia and flagella, and that perhaps a DNA virus led to the cell nucleus, though none of them are widely accepted.

Archaeans, bacteria, and eukaryotes continued to diversify and to become more complex and better adapted to their environments. Each domain repeatedly split into multiple lineages, although little is known about the history of the archaea and bacteria. Around 1.1 Ga, the supercontinent Rodinia was assembling. The plant, animal, and fungi lines had split, though they still existed as solitary cells. Some of these lived in colonies, and gradually a division of labor began to take place; for instance, cells on the periphery might have started to assume different roles from those in the interior. Although the division between a colony with specialized cells and a multicellular organism is not always clear, around 1 billion years ago the first multicellular plants emerged, probably green algae. Possibly by around 900 Ma true multicellularity had also evolved in animals.

At first it probably resembled today's sponges, which have totipotent cells that allow a disrupted organism to reassemble itself. As the division of labor was completed in all lines of multicellular organisms, cells became more specialized and more dependent on each other; isolated cells would die.

Supercontinents in the Proterozoic

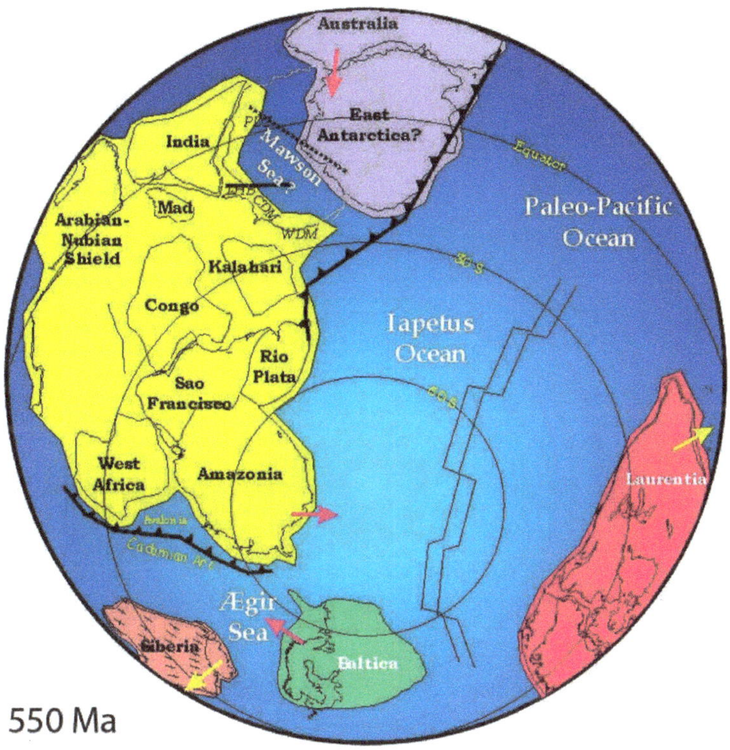

A reconstruction of Pannotia (550 Ma).

Reconstructions of tectonic plate movement in the past 250 million years (the Cenozoic and Mesozoic eras) can be made reliably using fitting of continental margins, ocean floor magnetic anomalies and paleomagnetic poles. No ocean crust dates back further than that, so earlier reconstructions are more difficult. Paleomagnetic poles are supplemented by geologic evidence such as orogenic belts, which mark the edges of ancient plates, and past distributions of flora and fauna. The further back in time, the scarcer and harder to interpret the data get and the more diverse the reconstructions.

Throughout the history of the Earth, there have been times when continents collided and formed a supercontinent, which later broke up into new continents. About 1000 to 830 Ma, most continental mass was united in the supercontinent Rodinia. Rodinia may have been preceded by Early-Middle Proterozoic continents called Nuna and Columbia.

After the break-up of Rodinia about 800 Ma, the continents may have formed another short-lived supercontinent around 550 Ma. The hypothetical supercontinent is sometimes referred to as Pannotia or Vendia. The evidence for it is a phase of continental collision known as the Pan-African orogeny, which joined the continental masses of current-day

Africa, South America, Antarctica and Australia. The existence of Pannotia depends on the timing of the rifting between Gondwana (which included most of the landmass now in the Southern Hemisphere, as well as the Arabian Peninsula and the Indian subcontinent) and Laurentia (roughly equivalent to current-day North America). It is at least certain that by the end of the Proterozoic eon, most of the continental mass lay united in a position around the south pole.

Late Proterozoic Climate and Life

A 580 million year old fossil of *Spriggina floundensi*, an animal from the Ediacaran period. Such life forms could have been ancestors to the many new forms that originated in the Cambrian Explosion.

The end of the Proterozoic saw at least two Snowball Earths, so severe that the surface of the oceans may have been completely frozen. This happened about 716.5 and 635 Ma, in the Cryogenian period. The intensity and mechanism of both glaciations are still under investigation and harder to explain than the early Proterozoic Snowball Earth. Most paleoclimatologists think the cold episodes were linked to the formation of the supercontinent Rodinia. Because Rodinia was centered on the equator, rates of chemical weathering increased and carbon dioxide (CO_2) was taken from the atmosphere. Because CO_2 is an important greenhouse gas, climates cooled globally. In the same way, during the Snowball Earths most of the continental surface was covered with permafrost, which decreased chemical weathering again, leading to the end of the glaciations. An alternative hypothesis is that enough carbon dioxide escaped through volcanic outgassing that the resulting greenhouse effect raised global temperatures.Increased volcanic activity resulted from the break-up of Rodinia at about the same time.

The Cryogenian period was followed by the Ediacaran period, which was characterized by a rapid development of new multicellular lifeforms. Whether there is a connection between the end of the severe ice ages and the increase in diversity of life is not clear, but it does not seem coincidental. The new forms of life, called Ediacara biota, were larger and more diverse than ever. Though the taxonomy of most Ediacaran life forms is unclear, some were ancestors of groups of modern life. Important developments were the origin of muscular and neural cells. None of the Ediacaran fossils had hard body parts like skeletons. These first appear after the boundary between the Proterozoic and Phanerozoic eons or Ediacaran and Cambrian periods.

PHANEROZOIC EON

The Phanerozoic is the current eon on Earth, which started approximately 542 million years ago. It consists of three eras: The Paleozoic, Mesozoic, and Cenozoic, and is the time when multi-cellular life greatly diversified into almost all the organisms known today.

The Paleozoic ("old life") era was the first and longest era of the Phanerozoic eon, lasting from 542 to 251 Ma. During the Paleozoic, many modern groups of life came into existence. Life colonized the land, first plants, then animals. Two major extinctions occurred. The continents formed at the break-up of Pannotia and Rodinia at the end of the Proterozoic slowly moved together again, forming the supercontinent Pangaea in the late Paleozoic.

The Mesozoic (middle life) era lasted from 251 Ma to 66 Ma. It is subdivided into the Triassic, Jurassic, and Cretaceous periods. The era began with the Permian–Triassic extinction event, the most severe extinction event in the fossil record; 95% of the species on Earth died out. It ended with the Cretaceous–Paleogene extinction event that wiped out the dinosaurs..

The Cenozoic (new life) era began at 66 Ma, and is subdivided into the Paleogene, Neogene, and Quaternary periods. These three periods are further split into seven subdivisions, with the Paleogene composed of The Paleocene, Eocene, and Oligocene, the Neocene divided into the Miocene, Pliocene, and the Quaternary composed of the Pleistocene, and Holocene. Mammals, birds, amphibians, crocodilians, turtles and lepidosaurs were able to survive the Cretaceous–Paleogene extinction event that killed off the non-avian dinosaurs and many other forms of life, and this is the era during which they diversified into their modern forms.

Tectonics, Paleogeography and Climate

At the end of the Proterozoic, the supercontinent Pannotia had broken apart into the smaller continents Laurentia, Baltica, Siberia and Gondwana. During periods when continents move apart, more oceanic crust is formed by volcanic activity. Because young volcanic crust is relatively hotter and less dense than old oceanic crust, the ocean floors rise during such periods. This causes the sea level to rise. Therefore, in the first half of the Paleozoic, large areas of the continents were below sea level.

Early Paleozoic climates were warmer than today, but the end of the Ordovician saw a short ice age during which glaciers covered the south pole, where the huge continent Gondwana was situated. Traces of glaciation from this period are only found on former Gondwana. During the Late Ordovician ice age, a few mass extinctions took place, in which many brachiopods, trilobites, Bryozoa and corals disappeared. These marine species could probably not contend with the decreasing temperature of the sea water. After the extinctions new species evolved, more diverse and better adapted. They would fill the niches left by the extinct species.

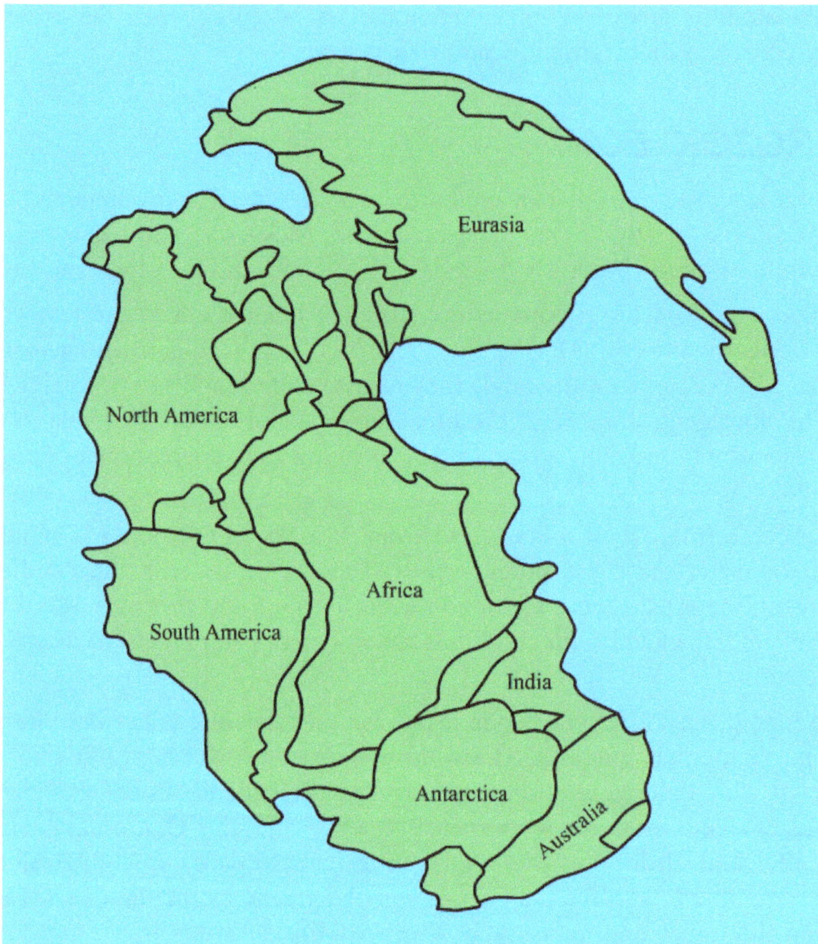

Pangaea was a supercontinent that existed from about 300 to 180 Ma. The outlines of the modern continents and other landmasses are indicated on this map.

The continents Laurentia and Baltica collided between 450 and 400 Ma, during the Caledonian Orogeny, to form Laurussia (also known as Euramerica). Traces of the mountain belt this collision caused can be found in Scandinavia, Scotland, and the northern Appalachians. In the Devonian period (416–359 Ma) Gondwana and Siberia began to move towards Laurussia. The collision of Siberia with Laurussia caused the Uralian Orogeny, the collision of Gondwana with Laurussia is called the Variscan or Hercynian Orogeny in Europe or the Alleghenian Orogeny in North America. The latter phase took place during the Carboniferous period (359–299 Ma) and resulted in the formation of the last supercontinent, Pangaea.

By 180 Ma, Pangaea broke up into Laurasia and Gondwana.

Cambrian Explosion

Trilobites first appeared during the Cambrian period and were among the most widespread and diverse groups of Paleozoic organisms.

The rate of the evolution of life as recorded by fossils accelerated in the Cambrian period (542–488 Ma). The sudden emergence of many new species, phyla, and forms in this period is called the Cambrian Explosion. The biological fomenting in the Cambrian Explosion was unpreceded before and since that time. Whereas the Ediacaran life forms appear yet primitive and not easy to put in any modern group, at the end of the Cambrian most modern phyla were already present. The development of hard body parts such as shells, skeletons or exoskeletons in animals like molluscs, echinoderms, crinoids and arthropods (a well-known group of arthropods from the lower Paleozoic are the trilobites) made the preservation and fossilization of such life forms easier than those of their Proterozoic ancestors. For this reason, much more is known about life in and after the Cambrian than about that of older periods. Some of these Cambrian groups appear complex but are quite different from modern life; examples are *Anomalocaris* and *Haikouichthys*.

During the Cambrian, the first vertebrate animals, among them the first fishes, had appeared. A creature that could have been the ancestor of the fishes, or was probably closely related to it, was *Pikaia*. It had a primitive notochord, a structure that could have developed into a vertebral column later. The first fishes with jaws (Gnathostomata) appeared during the next geological period, the Ordovician. The colonisation of new niches resulted in massive body sizes. In this way, fishes with increasing sizes evolved during the early Paleozoic, such as the titanic placoderm *Dunkleosteus*, which could grow

7 meters (23 ft) long. The diversity of life forms did not increase greatly because of a series of mass extinctions that define widespread biostratigraphic units called *biomeres*. After each extinction pulse, the continental shelf regions were repopulated by similar life forms that may have been evolving slowly elsewhere. By the late Cambrian, the trilobites had reached their greatest diversity and dominated nearly all fossil assemblages.

Colonization of Land

Artist's conception of Devonian flora.

Oxygen accumulation from photosynthesis resulted in the formation of an ozone layer that absorbed much of the Sun's ultraviolet radiation, meaning unicellular organisms that reached land were less likely to die, and prokaryotes began to multiply and become better adapted to survival out of the water. Prokaryote lineages had probably colonized the land as early as 2.6 Ga even before the origin of the eukaryotes. For a long time, the land remained barren of multicellular organisms. The supercontinent Pannotia formed around 600 Ma and then broke apart a short 50 million years later. Fish, the earliest vertebrates, evolved in the oceans around 530 Ma. A major extinction event occurred near the end of the Cambrian period, which ended 488 Ma.

Several hundred million years ago, plants (probably resembling algae) and fungi started growing at the edges of the water, and then out of it. The oldest fossils of land fungi and plants date to 480–460 Ma, though molecular evidence suggests the fungi may have colonized the land as early as 1000 Ma and the plants 700 Ma. Initially remaining close to the water's edge, mutations and variations resulted in further colonization of this new environment. The timing of the first animals to leave the oceans is not precisely known: the oldest clear evidence is of arthropods on land around 450 Ma, perhaps thriving and becoming better adapted due to the vast food source provided by the terrestrial plants. There is also unconfirmed evidence that arthropods may have appeared on land as early as 530 Ma.

Evolution of Tetrapods

Tiktaalik, a fish with limb-like fins and a predecessor of tetrapods. Reconstruction from fossils about 375 million years old.

At the end of the Ordovician period, 443 Ma, additional extinction events occurred, perhaps due to a concurrent ice age. Around 380 to 375 Ma, the first tetrapods evolved from fish. Fins evolved to become limbs that the first tetrapods used to lift their heads out of the water to breathe air. This would let them live in oxygen-poor water, or pursue small prey in shallow water. They may have later ventured on land for brief periods. Eventually, some of them became so well adapted to terrestrial life that they spent their adult lives on land, although they hatched in the water and returned to lay their eggs. This was the origin of the amphibians. About 365 Ma, another period of extinction occurred, perhaps as a result of global cooling. Plants evolved seeds, which dramatically accelerated their spread on land, around this time (by approximately 360 Ma).

About 20 million years later (340 Mathe amniotic egg evolved, which could be laid on land, giving a survival advantage to tetrapod embryos. This resulted in the divergence of amniotes from amphibians. Another 30 million years (310 Ma) saw the divergence of the synapsids (including mammals) from the sauropsids (including birds and reptiles). Other groups of organisms continued to evolve, and lines diverged—in fish, insects, bacteria, and so on—but less is known of the details.

Dinosaurs were the dominant terrestrial vertebrates throughout most of the Mesozoic.

After yet another, the most severe extinction of the period (251~250 Ma), around 230 Ma, dinosaurs split off from their reptilian ancestors. The Triassic–Jurassic extinction event at 200 Ma spared many of the dinosaurs, and they soon became dominant among the vertebrates. Though some mammalian lines began to separate during this period, existing mammals were probably small animals resembling shrews.

The boundary between avian and non-avian dinosaurs is not clear, but *Archaeopteryx*, traditionally considered one of the first birds, lived around 150 Ma.

The earliest evidence for the angiosperms evolving flowers is during the Cretaceous period, some 20 million years later (132 Ma).

Extinctions

The first mass extinction was the Ordovician-Silurian extinction. A possible cause was the intense glaciation of Gondwana, which eventually led to a snowball earth. 60% of marine invertebrates became extinct and 25% of all families.

The second mass extinction was the Late Devonian extinction, probably caused by the evolution of trees, which could lead to the depletion of greenhouse gases (like CO2) or lead to eutrophication of the water. 70% of all species became extinct.

The third mass extinction was the Permian-Triassic, or the Great Dying, event was possibly caused by some combination of the Siberian Traps volcanic event, an asteroid impact, methane hydrate gasification, sea level fluctuations, and a major anoxic event. Either the proposed Wilkes Land crater in Antarctica or Bedout structure off the northwest coast of Australia may indicate an impact connection with the Permian-Triassic extinction. But it remains uncertain whether either these or other proposed Permian-Triassic boundary craters are either real impact craters or even contemporaneous with the Permian-Triassic extinction event. This was by far the deadliest extinction ever, with about 57% of all families and 83% of all genera killed.

The fourth mass extinction was the Triassic-Jurassic extinction event in which almost all synapsids and archosaurs became extinct, probably due to new competition from dinosaurs.

The fifth and most recent mass extinction was the K-T extinction. In 66 Ma, a 10-kilometer (6.2 mi) asteroid struck Earth just off the Yucatán Peninsula - somewhere in the south western tip of then Laurasia - where the Chicxulub crater is today. This ejected vast quantities of particulate matter and vapor into the air that occluded sunlight, inhibiting photosynthesis. 75% of all life, including the non-avian dinosaurs, became extinct, marking the end of the Cretaceous period and Mesozoic era.

Diversification of Mammals

The first true mammals evolved in the shadows of dinosaurs and other large archosaurs that filled the world by the late Triassic. The first mammals were very small, and were probably nocturnal to escape predation. Mammal diversification truly began only after the Cretaceous-Paleogene extinction event. By the early Paleocene the earth recovered from the extinction, and mammalian diversity increased. Creatures like *Ambulocetus* took to the oceans to eventually evolve into whales, whereas some creatures, like primates, took to the trees. This all changed during the mid to late Eocene when the circum-Antarctic current formed between Antarctica and Australia which disrupted weather patterns on a global scale. Grassless savannas began to predominate much of the landscape, and mammals such as *Andrewsarchus* rose up to become the largest known terrestrial predatory mammal ever, and early whales like *Basilosaurus* took control of the seas.

The evolution of grass brought a remarkable change to the Earth's landscape, and the new open spaces created pushed mammals to get bigger and bigger. Grass started to expand in the Miocene, and the Miocene is where many modern- day mammals first appeared. Giant ungulates like *Paraceratherium* and *Deinotherium* evolved to rule the grasslands. The evolution of grass also brought primates down from the trees, and started human evolution. The first big cats evolved during this time as well. The Tethys Sea was closed off by the collision of Africa and Europe.

The formation of Panama was perhaps the most important geological event to occur in the last 60 million years. Atlantic and Pacific currents were closed off from each other, which caused the formation of the Gulf Stream, which made Europe warmer. The land bridge allowed the isolated creatures of South America to migrate over to North America, and vice versa. Various species migrated south, leading to the presence in South America of llamas, the spectacled bear, kinkajous and jaguars.

Three million years ago saw the start of the Pleistocene epoch, which featured dramatic climactic changes due to the ice ages. The ice ages led to the evolution of modern man in Saharan Africa and expansion. The mega-fauna that dominated fed on grasslands that, by now, had taken over much of the subtropical world. The large amounts of water held in the ice allowed for various bodies of water to shrink and sometimes disappear such as the North Sea and the Bering Strait. It is believed by many that a huge migration took place along Beringia which is why, today, there are camels (which evolved and became extinct in North America), horses (which evolved and became extinct in North America), and Native Americans. The ending of the last ice age coincided with the expansion of man, along with a massive die out of ice age mega-fauna. This extinction, nicknamed "the Sixth Extinction, has been going ever since.

Human Evolution

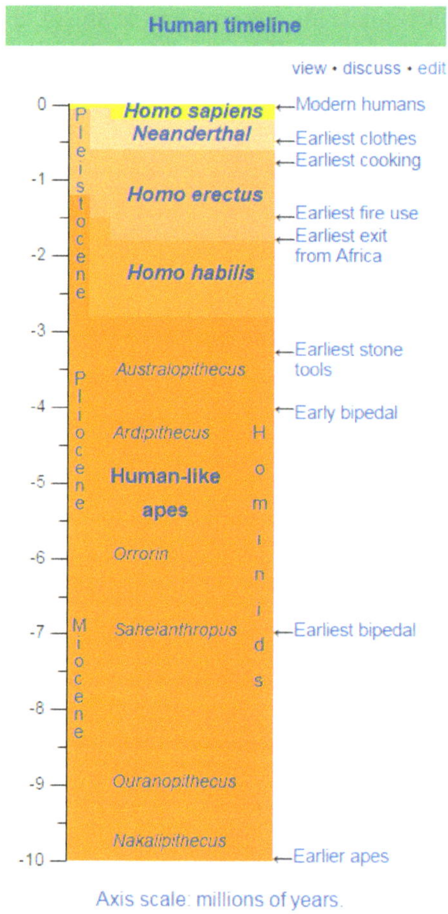

A small African ape living around 6 Ma was the last animal whose descendants would include both modern humans and their closest relatives, the chimpanzees. Only two branches of its family tree have surviving descendants. Very soon after the split, for reasons that are still unclear, apes in one branch developed the ability to walk upright. Brain size increased rapidly, and by 2 Ma, the first animals classified in the genus *Homo* had appeared. Of course, the line between different species or even genera is somewhat arbitrary as organisms continuously change over generations. Around the same time, the other branch split into the ancestors of the common chimpanzee and the ancestors of the bonobo as evolution continued simultaneously in all life forms.

The ability to control fire probably began in *Homo erectus* (or *Homo ergaster*), probably at least 790,000 years ago but perhaps as early as 1.5 Ma. The use and discovery of controlled fire may even predate *Homo erectus*. Fire was possibly used by the early

Lower Paleolithic (Oldowan) hominid *Homo habilis* or strong australopithecines such as *Paranthropus*.

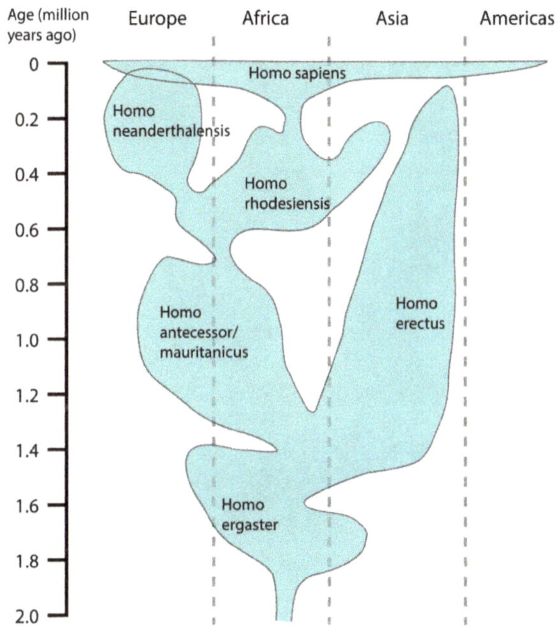

A reconstruction of human history based on fossil data.

It is more difficult to establish the origin of language; it is unclear whether *Homo erectus* could speak or if that capability had not begun until *Homo sapiens*. As brain size increased, babies were born earlier, before their heads grew too large to pass through the pelvis. As a result, they exhibited more plasticity, and thus possessed an increased capacity to learn and required a longer period of dependence. Social skills became more complex, language became more sophisticated, and tools became more elaborate. This contributed to further cooperation and intellectual development.Modern humans (*Homo sapiens*) are believed to have originated around 200,000 years ago or earlier in Africa; the oldest fossils date back to around 160,000 years ago.

The first humans to show signs of spirituality are the Neanderthals (usually classified as a separate species with no surviving descendants); they buried their dead, often with no sign of food or tools. However, evidence of more sophisticated beliefs, such as the early Cro-Magnon cave paintings (probably with magical or religious significance)[-19] did not appear until 32,000 years ago.Cro-Magnons also left behind stone figurines such as Venus of Willendorf, probably also signifying religious belief.[-19] By 11,000 years ago, *Homo sapiens* had reached the southern tip of South America, the last of the uninhabited continents (except for Antarctica, which remained undiscovered until 1820 AD). Tool use and communication continued to improve, and interpersonal relationships became more intricate.

Civilization

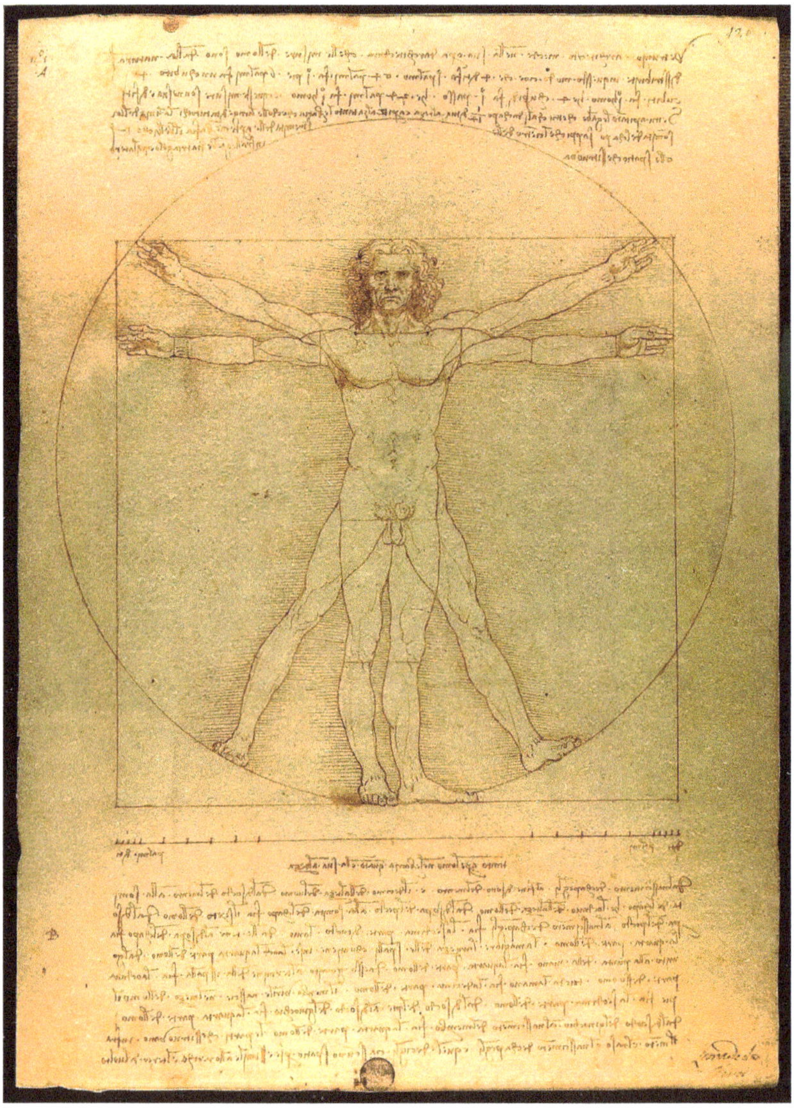

Vitruvian Man by Leonardo da Vinci epitomizes the advances in art and science seen during the Renaissance.

Throughout more than 90% of its history, *Homo sapiens* lived in small bands as nomadic hunter-gatherers. As language became more complex, the ability to remember and communicate information resulted, according to a theory proposed by Richard Dawkins, in a new replicator: the meme. Ideas could be exchanged quickly and passed down the generations. Cultural evolution quickly outpaced biological evolution, and history proper began. Between 8500 and 7000 BC, humans in the Fertile Crescent in the Middle East began the systematic husbandry of plants and animals: agriculture. This spread to neighboring regions, and developed independently elsewhere, until most *Homo sapiens* lived sedentary lives in permanent settlements as farmers. Not all societies abandoned nomadism, especially those in isolated areas of the globe poor in domesticable plant species, such as Australia. However, among those civilizations that did adopt agriculture, the relative stability and increased productivity provided by farming allowed the population to expand.

Agriculture had a major impact; humans began to affect the environment as never before. Surplus food allowed a priestly or governing class to arise, followed by increasing division of labor. This led to Earth's first civilization at Sumer in the Middle East, between 4000 and 3000 BC. Additional civilizations quickly arose in ancient Egypt, at the Indus River valley and in China. The invention of writing enabled complex societies to arise: record-keeping and libraries served as a storehouse of knowledge and increased the cultural transmission of information. Humans no longer had to spend all their time working for survival, enabling the first specialized occupations (e.g. craftsmen, merchants, priests, etc...). Curiosity and education drove the pursuit of knowledge and wisdom, and various disciplines, including science (in a primitive form), arose. This in turn led to the emergence of increasingly larger and more complex civilizations, such as the first empires, which at times traded with one another, or fought for territory and resources.

By around 500 BC, there were advanced civilizations in the Middle East, Iran, India, China, and Greece, at times expanding, at times entering into decline. In 221 BC, China became a single polity that would grow to spread its culture throughout East Asia, and it has remained the most populous nation in the world. The fundamentals of Western civilization were largely shaped in Ancient Greece, with the world's first democratic government and major advances in philosophy, science, and mathematics, and in Ancient Rome in law, government, and engineering. The Roman Empire was Christianized by Emperor Constantine in the early 4th century and declined by the end of the 5th. Beginning with the 7th century, Christianization of Europe began. In 610, Islam was founded and quickly became the dominant religion in Western Asia. In 1054 AD the Great Schism between the Roman Catholic Church and the Eastern Orthodox Church led to the prominent cultural differences between Western and Eastern Europe.

In the 14th century, the Renaissance began in Italy with advances in religion, art, and science.[-319] At that time the Christian Church as a political entity lost much of its power. In 1492, Christopher Columbus reached the Americas, initiating great changes to the new world. European civilization began to change beginning in 1500, leading to the scientific and industrial revolutions. That continent began to exert political and cultural dominance over human societies around the world, a time known as the Colonial era (also see Age of Discovery).[-299] In the 18th century a cultural movement known as the Age of Enlightenment further shaped the mentality of Europe and contributed to its secularization. From 1914 to 1918 and 1939 to 1945, nations around the world were embroiled in world wars. Established following World War I, the League of Nations was a first step in establishing international institutions to settle disputes peacefully. After failing to prevent World War II, mankind's bloodiest conflict, it was replaced by the United Nations. After the war, many new states were formed, declaring or being granted independence in a period of decolonization. The United States and Soviet Union became the world's dominant superpowers for a time, and they held an often-violent rivalry known as the Cold War until the dissolution of the latter. In 1992, several European nations joined in the European Union. As transportation and communication improved, the economies and political affairs of nations around the world have become increasingly intertwined. This globalization has often produced both conflict and cooperation.

Recent events

Astronaut Bruce McCandless II outside of the space shuttle *Challenger* in 1984.

Change has continued at a rapid pace from the mid-1940s to today. Technological developments include nuclear weapons, computers, genetic engineering, and nanotechnology. Economic globalization, spurred by advances in communication and transportation technology, has influenced everyday life in many parts of the world. Cultural and institutional forms such as democracy, capitalism, and environmentalism have increased influence. Major concerns and problems such as disease, war, poverty, violent radicalism, and recently, human-caused climate change have risen as the world population increases.

In 1957, the Soviet Union launched the first artificial satellite into orbit and, soon afterward, Yuri Gagarin became the first human in space. Neil Armstrong, an American, was the first to set foot on another astronomical object, the Moon. Unmanned probes have been sent to all the known planets in the solar system, with some (such as Voyager) having left the solar system. The Soviet Union and the United States were the earliest leaders in space exploration in the 20th century. Five space agencies, representing over fifteen countries, have worked together to build the International Space Station. Aboard it, there has been a continuous human presence in space since 2000. The World Wide Web became a part of everyday life in the 1990s, and since then has become an indispensable source of information in the developed world.

REFERENCES

1. Gradstein, Ogg & Smith 2004
2. "Age of the Earth". U.S. Geological Survey. 1997. Archived from the original on 23 December 2005. Retrieved 2006-01-10.
3. Dalrymple, G. Brent (2001). The age of the Earth in the twentieth century: a problem (mostly) solved. Special Publications, Geological Society of London. 190 (1): 205–221. Bibcode1GSLSP.190..205D. doi.1144/GSL.SP.2001.190.01.14.
4. Manhesa, Gérard; Allègre, Claude J.; Dupréa, Bernard & Hamelin, Bruno (1980). Lead isotope study of basic-ultrabasic layered complexes: Speculations about the age of the earth and primitive mantle characteristics. Earth and Planetary Science Letters. 47 (3): 370–382. Bibcode0E&PSL..47..370M. doi.1016/0012-821X(80)90024-2.
5. International Stratigraphic Chart. International Commission on Stratigraphy
6. Schopf, J. William; Kudryavtsev, Anatoliy B.; Czaja, Andrew D.; Tripathi, Abhishek B. (5 October 2007). Evidence of Archean life: Stromatolites and microfossils. Precambrian Research. Amsterdam, the Netherlands: Elsevier. 158 (3–4): 141–155. doi.1016/j.precamres.2007.04.009. ISSN 0301-9268.
7. Schopf, J. William (29 June 2006). Fossil evidence of Archaean life. Philosophical Transactions of the Royal Society B. London: Royal Society. 361 (1470): 869–885. doi.1098/rstb.2006.1834. ISSN 0962-8436. PMC 1578735. PMID 16754604.
8. Borenstein, Seth (13 November 2013). Oldest fossil found: Meet your microbial mom. Excite. Yonkers, NY: Mindspark Interactive Network. Associated Press. Retrieved 2015-06-02.
9. Pearlman, Jonathan (13 November 2013). 'Oldest signs of life on Earth found'. The Daily Telegraph. London: Telegraph Media Group. Retrieved 2014-12-15.
10. Noffke, Nora; Christian, Daniel; Wacey, David; Hazen, Robert M. (16 November 2013). Microbially Induced Sedimentary Structures Recording an Ancient Ecosystem in the ca. 3.48 Billion-Year-Old Dresser Formation, Pilbara, Western Australia. Astrobiology. New Rochelle, NY: Mary Ann Liebert, Inc. 13 (12): 1103–1124. Bibcode3AsBio..13.1103N. doi.1089/ast.2013.1030. ISSN 1531-1074. PMC 3870916. PMID 24205812.
11. Ohtomo, Yoko; Kakegawa, Takeshi; Ishida, Akizumi; et al. (January 2014). Evidence for biogenic graphite in early Archaean Isua metasedimentary rocks. Nature Geoscience. London: Nature Publishing Group. 7 (1): 25–28. Bibcode4NatGe...7...25O. doi.1038/ngeo2025. ISSN 1752-0894.
12. Borenstein, Seth (19 October 2015). Hints of life on what was thought to be desolate early Earth. Excite. Yonkers, NY: Mindspark Interactive Network. Associated Press. Retrieved 2015-10-20.

13. Bell, Elizabeth A.; Boehnike, Patrick; Harrison, T. Mark; et al. (19 October 2015). Potentially biogenic carbon preserved in a 4.1 billion-year-old zircon (PDF). Proc. Natl. Acad. Sci. U.S.A. Washington, D.C.: National Academy of Sciences. 112: 201517557. Bibcode5PNAS..11214518B. doi.1073/pnas.1517557112. ISSN 1091-6490. PMC 4664351. PMID 26483481. Retrieved 2015-10-20.Early edition, published online before print.
14. Kunin, W.E.; Gaston, Kevin, eds. (31 December 1996). The Biology of Rarity: Causes and consequences of rare—common differences. ISBN 978-0-412-63380-5. Retrieved 26 May 2015.
15. Stearns, Beverly Peterson; Stearns, S. C.; Stearns, Stephen C. (2000). Watching, from the Edge of Extinction. Yale University Press. p. 1921. ISBN 978-0-300-08469-6. Retrieved 2014-12-27.
16. Novacek, Michael J. (8 November 2014). Prehistory's Brilliant Future. New York Times. Retrieved 2014-12-25.
17. G. Miller; Scott Spoolman (2012). Environmental Science - Biodiversity Is a Crucial Part of the Earth's Natural Capital. Cengage Learning. p. 62. ISBN 978-1-133-70787-5. Retrieved 2014-12-27.
18. Mora, C.; Tittensor, D.P.; Adl, S.; Simpson, A.G.; Worm, B. (23 August 2011). How many species are there on Earth and in the ocean?. PLOS Biology. 9: e1001127. doi.1371/journal.pbio.1001127. PMC 3160336. PMID 21886479.
19. Staff (2 May 2016). Researchers find that Earth may be home to 1 trillion species. National Science Foundation. Retrieved 6 May 2016.
20. Gradstein, Ogg & van Kranendonk 2008
21. Encrenaz, T. (2004). The solar system (3rd ed.). Berlin: Springer. p. 89. ISBN 978-3-540-00241-3.
22. Matson, John (July 7, 2010). Luminary Lineage: Did an Ancient Supernova Trigger the Solar System's Birth?. Scientific American. Retrieved 2012-04-13.
23. P. Goldreich; W. R. Ward (1973). The Formation of Planetesimals. Astrophysical Journal. 183: 1051–1062. Bibcode3ApJ...183.1051G. doi.1086/152291.
24. Newman, William L. (2007-07-09). Age of the Earth. Publications Services, USGS. Retrieved 2007-09-20.
25. Stassen, Chris (2005-09-10). The Age of the Earth. TalkOrigins Archive. Retrieved 2008-12-30.
26. Stassen, Chris (2005-09-10). The Age of the Earth. The TalkOrigins Archive. Retrieved 2007-09-20.
27. Yin, Qingzhu; Jacobsen, S. B.; Yamashita, K.; Blichert-Toft, J.; Télouk, P.; Albarède, F. (2002). A short timescale for terrestrial planet formation from Hf-W chronometry of meteorites. Nature. 418 (6901): 949–952. Bibcode2Natur.418..949Y. doi.1038/nature00995. PMID 12198540.
28. Kokubo, Eiichiro; Ida, Shigeru (2002). Formation of protoplanet systems and diversity of planetary systems. The Astrophysical Journal. 581 (1): 666–680. Bibcode2ApJ...581..666K. doi.1086/344105.

29. Charles Frankel, 1996, Volcanoes of the Solar System, Cambridge University Press, pp. 7–8, ISBN 978-0-521-47770-3

30. J.A. Jacobs (1953). The Earth's inner core. Nature. 172 (4372): 297–298. Bibcode3Natur.172..297J. doi.1038/172297a0.

31. van Hunen, J.; van den Berg, A.P. (2007). Plate tectonics on the early Earth: Limitations imposed by strength and buoyancy of subducted lithosphere. Lithos. 103 (1-2): 217–235. Bibcode8Litho.103..217V. doi.1016/j.lithos.2007.09.016.

32. Wilde, S. A.; Valley, J.W.; Peck, W.H. & Graham, C.M. (2001). Evidence from detrital zircons for the existence of continental crust and oceans on the Earth 4.4 Gyr ago(PDF). Nature. 409 (6817): 175–178. Bibcode1Natur.409..175W. doi.1038/35051550. PMID 11196637. Retrieved 2013-05-25.

33. Lindsey, Rebecca; David Morrison; Robert Simmon (March 1, 2006). Ancient crystals suggest earlier ocean. Earth Observatory. NASA. Retrieved April 18, 2012.

34. Cavosie, A. J.; Valley, J. W.; Wilde, S. A.; Edinburgh Ion Microprobe Facility (E.I.M.F.) (2005). Magmatic $\delta^{18}O$ in 4400-3900 Ma detrital zircons: A record of the alteration and recycling of crust in the Early Archean". Earth and Planetary Science Letters. 235 (3–4): 663–681. Bibcode5E&PSL.235..663C. doi.1016/j.epsl.2005.04.028.

35. Belbruno, E.; Gott, J. Richard III (2005). Where Did The Moon Come From?. The Astronomical Journal. 129 (3): 1724–1745. arXiv:astro-ph/0405372. Bibcode5AJ....129.1724B. doi.1086/427539.

36. Münker, Carsten; Jörg A. Pfänder; Stefan Weyer; Anette Büchl; Thorsten Kleine; Klaus Mezger (July 4, 2003). Evolution of Planetary Cores and the Earth-Moon System from Nb/Ta Systematics. Science. 301 (5629): 84–87. Bibcode3Sci...301...84M. doi.1126/science.1084662. PMID 12843390. Retrieved 2012-04-13.

37. Nield, Ted (2009). Moonwalk (PDF). Geoscientist. Geological Society of London. 18 (9): 8. Retrieved April 18, 2012.

38. Britt, Robert Roy (2002-07-24). New Insight into Earth's Early Bombardment. Space.com. Retrieved 2012-02-09.

39. Green, Jack (2011). Academic Aspects of Lunar Water Resources and Their Relevance to Lunar Protolife. International Journal of Molecular Sciences. 12 (9): 6051–6076. doi.3390/ijms12096051. PMC 3189768. PMID 22016644.

40. Taylor, Thomas N.; Edith L. Taylor; Michael Krings (2006). Paleobotany: the biology and evolution of fossil plants. Academic Press. p. 49. ISBN 978-0-12-373972-8.

41. Steenhuysen, Julie (May 21, 2009). Study turns back clock on origins of life on Earth. Reuters.com. Reuters. Retrieved May 21, 2009.

42. Space Topics: Pluto and Charon. The Planetary Society. Retrieved 6 April 2010.

43. Pluto: Overview. Solar System Exploration. National Aeronautics and Space Administration. Retrieved 19 April 2012.

44. Kleine, T.; Palme, H.; Mezger, K.; Halliday, A.N. (2005). Hf-W Chronometry of Lunar Metals and the Age and Early Differentiation of the Moon. Science. 310 (5754): 1671–1674. Bibcode5Sci...310.1671K. doi.1126/science.1118842. PMID 16308422.

45. Halliday, A.N. (2006). The Origin of the Earth; What's New?. Elements. 2. pp. 205–210. doi.2113/gselements.2.4.205.

46. Halliday, Alex N (November 28, 2008). A young Moon-forming giant impact at 70–110 million years accompanied by late-stage mixing, core formation and degassing of the Earth. Philosophical Transactions of the Royal Society A. Philosophical Transactions of the Royal Society. 366 (1883): 4163–4181. Bibcode8RSPTA.366.4163H. doi.1098/rsta.2008.0209. PMID 18826916.

47. Williams, David R. (2004-09-01). Earth Fact Sheet. NASA. Retrieved 2010-08-09.

48. High Energy Astrophysics Science Archive Research Center (HEASARC). StarChild Question of the Month for October 2001. NASA Goddard Space Flight Center. Retrieved 20 April 2012.

49. Canup, R.M.; Asphaug, E. (2001). Origin of the Moon in a giant impact near the end of the Earth's formation. Nature. 412 (6848): 708–712. Bibcode1Natur.412..708C. doi.1038/35089010. PMID 11507633.

50. Liu, Lin-Gun (1992). Chemical composition of the Earth after the giant impact. Earth, Moon, and Planets. 57 (2): 85–97. Bibcode2EM&P...57...85L. doi.1007/BF00119610.

51. Newsom, Horton E.; Taylor, Stuart Ross (1989). Geochemical implications of the formation of the Moon by a single giant impact. Nature. 338 (6210): 29–34. Bibcode9Natur.338...29N. doi.1038/338029a0.

52. Taylor, G. Jeffrey (April 26, 2004). Origin of the Earth and Moon. NASA. Retrieved 2006-03-27., Taylor (2006) at the NASA website.

53. Davies, Geoffrey F. Mantle convection for geologists. Cambridge, UK: Cambridge University Press. ISBN 978-0-521-19800-4.

54. Cattermole, Peter; Moore, Patrick (1985). The story of the earth. Cambridge: Cambridge University Press. ISBN 978-0-521-26292-7.

55. Davies, Geoffrey F. (2011). Mantle convection for geologists. Cambridge, UK: Cambridge University Press. ISBN 978-0-521-19800-4.

56. Bleeker, W.; B. W. Davis (May 2004). What is a craton?. Spring meeting. American Geophysical Union. T41C-01.

57. ondie, Kent C. (1997). Plate tectonics and crustal evolution (4th ed.). Oxford: Butterworth Heinemann. ISBN 978-0-7506-3386-4.

58. Holland, Heinrich D. (June 2006). The oxygenation of the atmosphere and oceans. The Royal Society. doi.1098/rstb.2006.1838+Phil.+Trans.+R.+Soc.+B+29+June+2006+vol.+361+no.+1470+903-915. Retrieved 2010-02-17.

59. Kasting, James F. (1993). Earth's early atmosphere. Science. 259 (5097): 920–926. doi.1126/science.11536547. PMID 11536547.

60. Gale, Joseph (2009). Astrobiology of Earth : the emergence, evolution, and

future of life on a planet in turmoil. Oxford: Oxford University Press. ISBN 978-0-19-920580-6.

61. Kasting, James F.; Catling, David (2003). Evolution of a habitable planet. Annual Review of Astronomy and Astrophysics. 41 (1): 429–463. Bibcode3ARA&A..41..429K. doi.1146/annurev.astro.41.071601.170049.

62. Kasting, James F.; Howard, M. Tazewell (September 7, 2006). Atmospheric composition and climate on the early Earth (PDF). Philosophical Transactions of the Royal Society B. 361 (361): 1733–1742. doi.1098/rstb.2006.1902.

63. Selsis, Franck (2005). Chapter 11. The Prebiotic Atmosphere of the Earth. Astrobiology: Future perspectives. Astrophysics and space science library. 305. pp. 267–286. doi.1007/1-4020-2305-7_11.

64. Morbidelli, A.; Chambers, J.; Lunine, J. I.; Petit, J. M.; Robert, F.; Valsecchi, G. B.; Cyr, K. E. (2000). Source regions and timescales for the delivery of water to the Earth. Meteoritics & Planetary Science. 35 (6): 1309–1320. Bibcode0M&PS...35.1309M. doi.1111/j.1945-5100.2000.tb01518.x.

65. The Sun›s evolution

66. Sagan, Carl; Mullen, George (July 7, 1972). Earth and Mars: Evolution of Atmospheres and Surface Temperatures. Science. 177 (4043): 52–56. Bibcode2Sci...177...52S. doi.1126/science.177.4043.52. PMID 17756316.

67. Szathmáry, E. (February 2005). In search of the simplest cell. Nature. 433 (7025): 469–470. Bibcode5Natur.433..469S. doi.1038/433469a. PMID 15690023. Retrieved 2008-09-01.

68. Luisi, P. L.; Ferri, F. & Stano, P. (2006). Approaches to semi-synthetic minimal cells: a review. Naturwissenschaften. 93 (1): 1–13. Bibcode6NW.....93....1L. doi.1007/s00114-005-0056-z. PMID 16292523.

69. A. Lazcano; J. L. Bada (June 2004). The 1953 Stanley L. Miller Experiment: Fifty Years of Prebiotic Organic Chemistry. Origins of Life and Evolution of Biospheres. 33 (3): 235–242. doi.1023/A4807125069. PMID 14515862.

70. Dreifus, Claudia (2010-05-17). A Conversation With Jeffrey L. Bada: A Marine Chemist Studies How Life Began. nytimes.com.

71. Moskowitz, Clara (29 March 2012). Life's Building Blocks May Have Formed in Dust Around Young Sun. Space.com. Retrieved 30 March 2012.

72. Peretó, J. (2005). Controversies on the origin of life(PDF). Int. Microbiol. 8 (1): 23–31. PMID 15906258. Retrieved 2007-10-07.

73. Joyce, G.F. (2002). The antiquity of RNA-based evolution. Nature. 418 (6894): 214–21. Bibcode2Natur.418..214J. doi.1038/418214a. PMID 12110897.

74. Hoenigsberg, H. (December 2003). Evolution without speciation but with selection: LUCA, the Last Universal Common Ancestor in Gilbert's RNA world. Genetic and Molecular Research. 2 (4): 366–375. PMID 15011140. Retrieved 2008-08-30.(also available as PDF)

75. Forterre, Patrick (2005). The two ages of the RNA world, and the transition to the DNA world: a story of viruses and cells. Biochimie. 87 (9–10): 793–803.

doi.1016/j.biochi.2005.03.015. PMID 16164990.

76. Cech, T.R. (August 2000). The ribosome is a ribozyme. Science. 289 (5481): 878–9. doi.1126/science.289.5481.878. PMID 10960319. Retrieved 2008-09-01.
77. Johnston WK, Unrau PJ, et al. (2001). RNA-Catalyzed RNA Polymerization: Accurate and General RNA-Templated Primer Extension. Science. 292 (5520): 1319–1325. Bibcode1Sci...292.1319J. doi.1126/science.1060786. PMID 11358999.
78. Levy, M. & Miller, S.L. (July 1998). The stability of the RNA bases: Implications for the origin of life. Proc. Natl. Acad. Sci. U.S.A. 95 (14): 7933–8. Bibcode8PNAS...95.7933L. doi.1073/pnas.95.14.7933. PMC 20907. PMID 9653118.
79. Larralde, R.; Robertson, M. P. & Miller, S. L. (August 1995). Rates of decomposition of ribose and other sugars: implications for chemical evolution. Proc. Natl. Acad. Sci. U.S.A. 92 (18): 8158–60. Bibcode5PNAS...92.8158L. doi.1073/pnas.92.18.8158. PMC 41115. PMID 7667262.
80. Lindahl, T. (April 1993). Instability and decay of the primary structure of DNA. Nature. 362 (6422): 709–15. Bibcode3Natur.362..709L. doi.1038/362709a0. PMID 8469282.
81. Orgel, L. (November 2000). A simpler nucleic acid. Science. 290 (5495): 1306–7. doi.1126/science.290.5495.1306. PMID 11185405.
82. Nelson, K.E.; Levy, M. & Miller, S.L. (April 2000). Peptide nucleic acids rather than RNA may have been the first genetic molecule. Proc. Natl. Acad. Sci. U.S.A. 97 (8): 3868–71. Bibcode0PNAS...97.3868N. doi.1073/pnas.97.8.3868. PMC 18108. PMID 10760258.
83. Dawkins, Richard (1996) [1986]. Origins and miracles. The Blind Watchmaker. New York: W. W. Norton & Company. ISBN 978-0-393-31570-7.
84. Davies, Paul (October 6, 2005). A quantum recipe for life. Nature. 437 (7060): 819. Bibcode5Natur.437..819D. doi.1038/437819a. PMID 16208350.(subscription required)
85. Martin, W. & Russell, M.J. (2003). On the origins of cells: a hypothesis for the evolutionary transitions from abiotic geochemistry to chemoautotrophic prokaryotes, and from prokaryotes to nucleated cells. Philosophical Transactions of the Royal Society B. 358 (1429): 59–85. doi.1098/rstb.2002.1183. PMC 1693102. PMID 12594918.
86. Kauffman, Stuart A. (1993). The origins of order : self-organization and selection in evolution (Reprint ed.). New York: Oxford University Press. ISBN 978-0-19-507951-7.
87. Wächtershäuser, G. (August 2000). Life as we don't know it. Science. 289 (5483): 1307–8. doi.1126/science.289.5483.1307. PMID 10979855.
88. Vasas, V.; Szathmáry, E.; Santos, M. (4 January 2010). Lack of evolvability in self-sustaining autocatalytic networks constraints metabolism-first scenarios for the origin of life. Proceedings of the National Academy of Sciences. 107 (4): 1470–1475. Bibcode0PNAS..107.1470V. doi.1073/pnas.0912628107.

89. Trevors, J.T. & Psenner, R. (2001). From self-assembly of life to present-day bacteria: a possible role for nanocells. FEMS Microbiol. Rev. 25 (5): 573–82. doi.1111/j.1574-6976.2001.tb00592.x. PMID 11742692.

90. Segré, D.; Ben-Eli, D.; Deamer, D. & Lancet, D. (February–April 2001). The Lipid World (PDF). Origins of Life and Evolution of Biospheres 2001. 31 (1–2): 119–45. doi.1023/A6746807104. PMID 11296516. Retrieved 2008-09-01.

91. Cairns-Smith, A.G. (1968). An approach to a blueprint for a primitive organism. In Waddington, C.H. Towards a Theoretical Biology. 1. Edinburgh University Press. pp. 57–66.

92. Ferris, J.P. (June 1999). Prebiotic Synthesis on Minerals: Bridging the Prebiotic and RNA Worlds. Biological Bulletin. Evolution: A Molecular Point of View. 196 (3): 311–314. doi.2307/1542957. JSTOR 1542957. PMID 10390828.

93. Hanczyc, M.M.; Fujikawa, S.M. & Szostak, Jack W. (October 2003). Experimental Models of Primitive Cellular Compartments: Encapsulation, Growth, and Division. Science. 302 (5645): 618–622. Bibcode3Sci...302..618H. doi.1126/science.1089904. PMID 14576428. Retrieved 2008-09-01.

94. Hartman, H. (October 1998). Photosynthesis and the Origin of Life. Origins of Life and Evolution of Biospheres. 28 (4–6): 512–521. Bibcode8OLEB...28..515H. doi.1023/A6548904157. PMID 11536891. Retrieved 2008-09-01.

95. Penny, David; Poole, Anthony (December 1999). The nature of the last universal common ancestor. Current Opinion in Genetics & Development. 9 (6): 672–677. doi.1016/S0959-437X(99)00020-9. PMID 10607605. Archived from the original on February 25, 2010. (PDF)

96. Earliest Life. University of Münster. 2003. Retrieved 2006-03-28.

97. Condie, Kent C. Earth as an Evolving Planetary System (2nd ed.). Burlington: Elsevier Science. ISBN 978-0-12-385228-1.

98. Leslie, M. (2009). On the Origin of Photosynthesis. Science. 323 (5919): 1286–1287. doi.1126/science.323.5919.1286. PMID 19264999.

99. Nisbet, E. G.; Sleep, N. H. (2001). The habitat and nature of early life. Nature. 409 (6823): 1083–1091. Bibcode1Natur.409.1083N. doi.1038/35059210. PMID 11234022.

100. De Marais, David J.; D (September 8, 2000). Evolution: When Did Photosynthesis Emerge on Earth?. Science. 289 (5485): 1703–1705. doi.1126/science.289.5485.1703. PMID 11001737.

101. Olson, John M. (February 2, 2006). Photosynthesis in the Archean Era. Photosynthesis Research. 88 (2 / May, 2006): 109–17. doi.1007/s11120-006-9040-5. PMID 16453059. Retrieved 2010-02-16.

102. Fortey, Richard (September 1999) [1997]. Dust to Life. Life: A Natural History of the First Four Billion Years of Life on Earth. New York: Vintage Books. ISBN 978-0-375-70261-7.

103. Chaisson, Eric J. (2005). Early Cells. Cosmic Evolution. Tufts University. Archived from the original on July 22, 2007. Retrieved 2006-03-29.

104. Snowball Earth. snowballearth.org. 2006–2009. Retrieved 2012-04-13.
105. What caused the snowball earths?. snowballearth.org. 2006–2009. Retrieved 2012-04-13.
106. Allaby, Michael, ed. (2013). Snowball Earth. Oxford Dictionary of Geology & Earth Sciences (4th ed.). Oxford University Press. p. 539. ISBN 978-0-19-965306-5.
107. Bjornerud, Marcia (2005). Reading the Rocks: The Autobiography of the Earth. Westview Press. pp. 131–38. ISBN 978-0-8133-4249-8.
108. Slushball Earth hypothesis. Encylopædia Britannica.
109. Woese, Carl; Gogarten, J. Peter (October 21, 1999). When did eukaryotic cells evolve? What do we know about how they evolved from earlier life-forms?. Scientific American. Retrieved 2012-04-13.
110. Andersson, Siv G. E.; Zomorodipour, Alireza; Andersson, Jan O.; Sicheritz-Pontén, Thomas; Alsmark, U. Cecilia M.; Podowski, Raf M.; Näslund, A. Kristina; Eriksson, Ann-Sofie; Winkler, Herbert H.; Kurland, Charles G. (November 12, 1998). The genome sequence of Rickettsia prowazekii and the origin of mitochondria. Nature. 396 (6707): 133–140. Bibcode8Natur.396..133A. doi.1038/24094. PMID 9823893.
111. From prokaryotes to eukaryotes. Understanding evolution: your one-stop source for information on evolution. University of California Museum of Paleontology. Retrieved 2012-04-16.
112. Berglsand, Kristin J.; Haselkorn, Robert (June 1991). Evolutionary Relationships among the Eubacteria, Cyanobacteria, and Chloroplasts: Evidence from the rpoC1Gene of Anabaena sp. Strain PCC 7120. Journal of Bacteriology. 173 (11): 3446–3455. PMC 207958. PMID 1904436. (PDF)
113. Takemura, Masaharu (May 2001). Poxviruses and the origin of the eukaryotic nucleus. Journal of Molecular Evolution. 52 (5): 419–425. doi.1007/s002390010171. PMID 11443345.
114. Bell, Philip J (September 2001). Viral eukaryogenesis: was the ancestor of the nucleus a complex DNA virus?. Journal of Molecular Evolution. 53 (3): 251–256. doi.1007/s002390010215. PMID 11523012.
115. Gabaldón, Toni; Berend Snel; Frank van Zimmeren; Wieger Hemrika; Henk Tabak; Martijn A. Huynen (March 23, 2006). Origin and evolution of the peroxisomal proteome (PDF). Biology Direct. 1 (1): 8. doi.1186/1745-6150-1-8. PMC 1472686. PMID 16556314.
116. Hanson RE, James L. Crowley, Samuel A. Bowring, Jahandar Ramezani, et al. (May 21, 2004). Coeval Large-Scale Magmatism in the Kalahari and Laurentian Cratons During Rodinia Assembly. Science. 304 (5674): 1126–1129. Bibcode4Sci...304.1126H. doi.1126/science.1096329. PMID 15105458. Retrieved 2012-04-13.
117. Li, Z.X.; Bogdanova, S.V.; Collins, A.S.; Davidson, A.; De Waele, B.; Ernst, R.E.; Fitzsimons, I.C.W.; Fuck, R.A.; Gladkochub, D.P.; Jacobs, J.; Karlstrom,

K.E.; Lu, S.; Natapov, L.M.; Pease, V.; Pisarevsky, S.A.; Thrane, K.; Vernikovsky, V. (2008). Assembly, configuration, and break-up history of Rodinia: A synthesis. Precambrian Research. 160 (1–2): 179–210. doi.1016/j.precamres.2007.04.021.

118. Chaisson, Eric J. (2005). Ancient Fossils. Cosmic Evolution. Tufts University. Archived from the original on July 22, 2007. Retrieved 2006-03-31.

119. Bhattacharya, Debashish; Medlin, Linda (1998). Algal Phylogeny and the Origin of Land Plants. Plant Physiology. 116 (1): 9–15. doi.1104/pp.116.1.9. (PDF)

120. Kearey, Philip; Keith A. Klepeis; Frederick J. Vine (2009). Global tectonics. (3rd ed.). Oxford: Wiley-Blackwell. ISBN 978-1-4051-0777-8.

121. Torsvik, T. H. (30 May 2003). The Rodinia Jigsaw Puzzle. Science. 300 (5624): 1379–1381. doi.1126/science.1083469. PMID 12775828.

122. Zhao, Guochun; Cawood, Peter A.; Wilde, Simon A.; Sun, M. (2002). Review of global 2.1–1.8 Ga orogens: implications for a pre-Rodinia supercontinent. Earth-Science Reviews. 59 (1–4): 125–162. Bibcode2ESRv...59..125Z. doi.1016/S0012-8252(02)00073-9.

123. Zhao, Guochun; Sun, M.; Wilde, Simon A.; Li, S.Z. (2004). A Paleo-Mesoproterozoic supercontinent: assembly, growth and breakup. Earth-Science Reviews. 67 (1–2): 91–123. Bibcode4ESRv...67...91Z. doi.1016/j.earscirev.2004.02.003.

124. McElhinny, Michael W.; Phillip L. McFadden (2000). Paleomagnetism continents and oceans (2nd ed.). San Diego: Academic Press. ISBN 978-0-12-483355-5.

125. Dalziel, I.W.D. (1995). Earth before Pangea. Scientific American. 1 (1): 58–63. Bibcode5SciAm.272a..58D. doi.1038/scientificamerican0195-58.

126. Snowball Earth: New Evidence Hints at Global Glaciation 716.5 Million Years Ago. Science Daily. Mar 4, 2010. Retrieved April 18, 2012.

127. Snowball Earth' Hypothesis Challenged. Retrieved 29 September 2012.

128. Hoffman, P.F.; Kaufman, A.J.; Halverson, G.P.; Schrag, D.P. (1998). A Neoproterozoic Snowball Earth. Science. 281 (5381): 1342–1346. Bibcode8Sci...281.1342H. doi.1126/science.281.5381.1342. PMID 9721097.

129. Two Explosive Evolutionary Events Shaped Early History Of Multicellular Life. Science Daily. Jan 3, 2008. Retrieved April 18, 2012.

130. Xiao, S.; Laflamme, M. (2009). On the eve of animal radiation: phylogeny, ecology and evolution of the Ediacara biota. Trends in Ecology and Evolution. 24 (1): 31–40. doi.1016/j.tree.2008.07.015. PMID 18952316.

131. Patwardhan, A.M. (2010). The Dynamic Earth System. New Delhi: PHI Learning Private Limited. p. 146. ISBN 978-81-203-4052-7.

132. The Day the Earth Nearly Died. Horizon. BBC. 2002. Retrieved 2006-04-09.

133. The Cenozoic Era. University of California Museum of Paleontology. June 2011. Retrieved 2016-01-10.

134. Pannotia. UCMP Glossary. Retrieved 2006-03-12.

135. The Mass Extinctions: The Late Ordovician Extinction. BBC. Archived from

the original on 2006-02-21. Retrieved 2006-05-22.

136. Murphy, Dennis C. (May 20, 2006). The paleocontinent Euramerica. Devonian Times. Retrieved April 18, 2012.

137. Runkel, Anthony C.; Mackey, Tyler J.; Cowan, Clinton A.; Fox, David L. (1 November 2010). Tropical shoreline ice in the late Cambrian: Implications for Earth's climate between the Cambrian Explosion and the Great Ordovician Biodiversification Event. GSA Today: 4–10. doi.1130/GSATG84A.1.

138. Palmer, Allison R. (1984). The biomere problem: Evolution of an idea. Journal of Paleontology. 58 (3): 599–611.

139. Hallam, A.; Wignall, P.B. (1997). Mass extinctions and their aftermath (Repr. ed.). Oxford [u.a.]: Oxford Univ. Press. ISBN 978-0-19-854916-1.

140. Battistuzzi, Fabia U.; Feijao, Andreia; Hedges, S. Blair (2004). A genomic timescale of prokaryote evolution: insights into the origin of methanogenesis, phototrophy, and the colonization of land. BMC Evolutionary Biology. 4: 44. doi.1186/1471-2148-4-44. PMC 533871. PMID 15535883.

141. Pisani, Davide; Poling, Laura L.; Lyons-Weiler, Maureen; Hedges, S. Blair (January 19, 2004). The colonization of land by animals: molecular phylogeny and divergence times among arthropods. BMC Biology. 2: 1. doi.1186/1741-7007-2-1. PMC 333434. PMID 14731304.

142. Lieberman, Bruce S. (2003). Taking the Pulse of the Cambrian Radiation. Integrative and Comparative Biology. 43 (1): 229–237. doi.1093/icb/43.1.229. PMID 21680426.

143. The Mass Extinctions: The Late Cambrian Extinction. BBC. Retrieved 2006-04-09.

144. Landing, E.; Bowring, S. A.; Davidek, K. L.; Fortey, R. A.; Wimbledon, W. A. P. (2000). Cambrian–Ordovician boundary age and duration of the lowest Ordovician Tremadoc Series based on U–Pb zircon dates from Avalonian Wales. Geological Magazine. 137 (5): 485–494. doi.1017/S0016756800004507. (abstract)

145. Fortey, Richard (September 1999) [1997]. Landwards, Humanity. Life: A Natural History of the First Four Billion Years of Life on Earth. New York: Vintage Books. pp. 138–140, 300. ISBN 978-0-375-70261-7.

146. Heckman, D. S.; D. M. Geiser; B. R. Eidell; R. L. Stauffer; N. L. Kardos; S. B. Hedges (August 10, 2001). Molecular evidence for the early colonization of land by fungi and plants. Science. 293 (5532): 1129–1133. doi.1126/science.1061457. PMID 11498589.(abstract)

147. Johnson, E. W.; D. E. G. Briggs; R. J. Suthren; J. L. Wright; S. P. Tunnicliff (1 May 1994). Non-marine arthropod traces from the subaereal Ordivician Borrowdale volcanic group, English Lake District. Geological Magazine. 131 (3): 395–406. doi.1017/S0016756800011146. Retrieved 2012-04-13. (abstract)

148. MacNaughton, Robert B.; Jennifer M. Cole; Robert W. Dalrymple; Simon J. Braddy; Derek E. G. Briggs; Terrence D. Lukie (2002). First steps on land: Arthropod trackways in Cambrian-Ordovician eolian sandstone, southeastern

Ontario, Canada. Geology. 30 (5): 391–394. Bibcode2Geo....30..391M. doi.1130/0091-7613(2002)030<0391:FSOLAT>2.0.CO;2. ISSN 0091-7613. (abstract)

149. Clack, Jennifer A. (December 2005). Getting a Leg Up on Land. Scientific American. 293 (6): 100–7. Bibcode5SciAm.293f.100C. doi.1038/scientificamerican1205-100. PMID 16323697. Retrieved 2012-04-13.

150. McGhee, Jr, George R. (1996). The Late Devonian Mass Extinction: the Frasnian/Famennian Crisis. Columbia University Press. ISBN 978-0-231-07504-6.

151. Willis, K. J.; J. C. McElwain (2002). The Evolution of Plants. Oxford: Oxford University Press. p. 93. ISBN 978-0-19-850065-0.

152. Plant Evolution. Evolution for teaching. University of Waikato. October 2004. Retrieved April 18, 2012.

153. Wright, Jo (1999). New Blood. Walking with Dinosaurs. Episode 1. BBC.

154. The Mass Extinctions: The Late Triassic Extinction. BBC. Archived from the original on 2006-08-13. Retrieved 2006-04-09.

155. Archaeopteryx: An Early Bird. University of California, Berkeley Museum of Paleontology. 1996. Retrieved 2006-04-09.

156. Soltis, Pam; Doug Soltis; Christine Edwards (2005). Angiosperms. The Tree of Life Project. Retrieved 2006-04-09.

157. Big crater seen beneath ice sheet. BBC News. 3 June 2006. Retrieved April 18, 2012.

158. Benton M J (2005). When life nearly died: the greatest mass extinction of all time. London: Thames & Hudson. ISBN 0-500-28573-X.

159. Carl T. Bergstrom; Lee Alan Dugatkin (2012). Evolution. Norton. p. 515. ISBN 978-0-393-92592-0.

160. Chaisson, Eric J. (2005). Recent Fossils. Cosmic Evolution. Tufts University. Archived from the original on July 22, 2007. Retrieved 2006-04-09.

161. Strauss, Bob. The First Mammals:The Early Mammals of the Triassic, Jurassic and Cretaceous Periods. about.com. Retrieved 12 May 2015.

162. A Walking Whale: Ambulocetus. American Museum of Natural History. 2014-05-01. Retrieved 2016-01-10.

163. O'Neil, Dennis (2012). Early Primate Evolution: The First Primates. Palomar College. Retrieved 2016-01-10.

164. Andrewsarchus, Superb Skull of a Gigantic Beast, Now on View in Whales Exhibit. American Museum of Natural History. 2014-05-01. Retrieved 2016-01-10.

165. George Dvorsky (November 13, 2013). The world's first big cats came from Asia, not Africa. Io9.com. Retrieved 2016-01-10.

166. Hamon, N.; Sepulchre, P.; Lefebvre, V.; Ramstein, G. (2013). The role of eastern Tethys seaway closure in the Middle Miocene Climatic Transition (ca. 14 Ma) (PDF). Climate of the Past (9): 2687–2702. Bibcode3CliPa...9.2687H. doi.5194/cp-9-2687-2013. Retrieved 2016-01-10.

167. N.A.S.A. Isthmus of Panama. N.A.S.A.

168. Goren-Inbar, Naama; Nira Alperson; Mordechai E. Kislev; Orit Simchoni; Yoel Melamed; Adi Ben-Nun; Ella Werker (2004-04-30). Evidence of Hominin Control of Fire at Gesher Benot Ya'aqov, Israel. Science. 304 (5671): 725–727. Bibcode4Sci...304..725G. doi.1126/science.1095443. PMID 15118160. Retrieved 2012-04-13. (abstract)

169. McClellan (2006). Science and Technology in World History: An Introduction. Baltimore, Maryland: JHU Press. ISBN 978-0-8018-8360-6.

170. Reed, David L.; Smith, Vincent S.; Hammond, Shaless L.; Rogers, Alan R.; et al. (2004). Genetic Analysis of Lice Supports Direct Contact between Modern and Archaic Humans. PLoS Biology. 2 (11): e340. doi.1371/journal.pbio.0020340. PMC 521174. PMID 15502871.

171. Gibbons, Ann (2003). Oldest Members of Homo sapiensDiscovered in Africa. Science. 300 (5626): 1641. doi.1126/science.300.5626.1641. PMID 12805512. Retrieved 2012-04-13. (abstract)

172. Hopfe, Lewis M. (1987) [1976]. Characteristics of Basic Religions. Religions of the World (4th ed.). New York: MacMillan Publishing Company. pp. 17, 17–19. ISBN 978-0-02-356930-2.

173. Chauvet Cave. Metropolitan Museum of Art. Retrieved 2006-04-11.

174. Patrick K. O'Brien, ed. (2003) [2002]. The Human Revolution. Atlas of World History (concise ed.). New York: Oxford University Press. p. 16. ISBN 978-0-19-521921-0.

175. Dawkins, Richard (1989) [1976]. Memes: the new replicators. The Selfish Gene (2nd ed.). Oxford: Oxford University Press. pp. 189–201. ISBN 978-0-19-286092-7.

176. Tudge, Colin (1998). Neanderthals, Bandits and Farmers: How Agriculture Really Began. London: Weidenfeld & Nicolson. ISBN 978-0-297-84258-3.

177. Diamond, Jared (1999). Guns, Germs, and Steel. W. W. Norton & Company. ISBN 978-0-393-31755-8.

178. Jonathan Daly (19 December 2013). The Rise of Western Power: A Comparative History of Western Civilization. A&C Black. pp. 7–9. ISBN 978-1-4411-1851-6.

179. Human Spaceflight and Exploration – European Participating States. ESA. 2006. Retrieved 2006-03-27.

180. Expedition 13: Science, Assembly Prep on Tap for Crew. NASA. January 11, 2006. Retrieved 2006-03-27.

CHAPTER 2

Theory of Plate Tectonics

Chapter Contents

- **Intrtoduction**
- **Key Principles**
- **Types of Plate Boundaries**
- **Driving Forces of Plate Motion**
- **Development of the Theory**
- **Implications for Biogeography**
- **Plate Reconstruction**
- **Current Plates**

Plate tectonics is the study of how the Earth's crust is shaped by geological forces. It relies on the understanding that the crust is divided into large pieces, or plates, that sit on the molten magma present below the surface. Currents within the interior cause the plates to move, which causes many different geological events, including earthquakes and the forming of mountains and volcanoes. Understanding how plates move and interact is the main purpose of plate tectonics.

INTRTODUCTION

Plate tectonics (from the Late Latin *tectonicus*, from the Greek: τεκτονικός "pertaining to building") is a scientific theory describing the large-scale motion of 7 large plates and the movements of a larger number of smaller plates of the Earth›s lithosphere, over the last hundreds of millions of years. The theoretical model builds on the concept of continental drift developed during the first few decades of the 20th century. The geoscientific community accepted plate-tectonic theory after seafloor spreading was validated in the late 1950s and early 1960s.

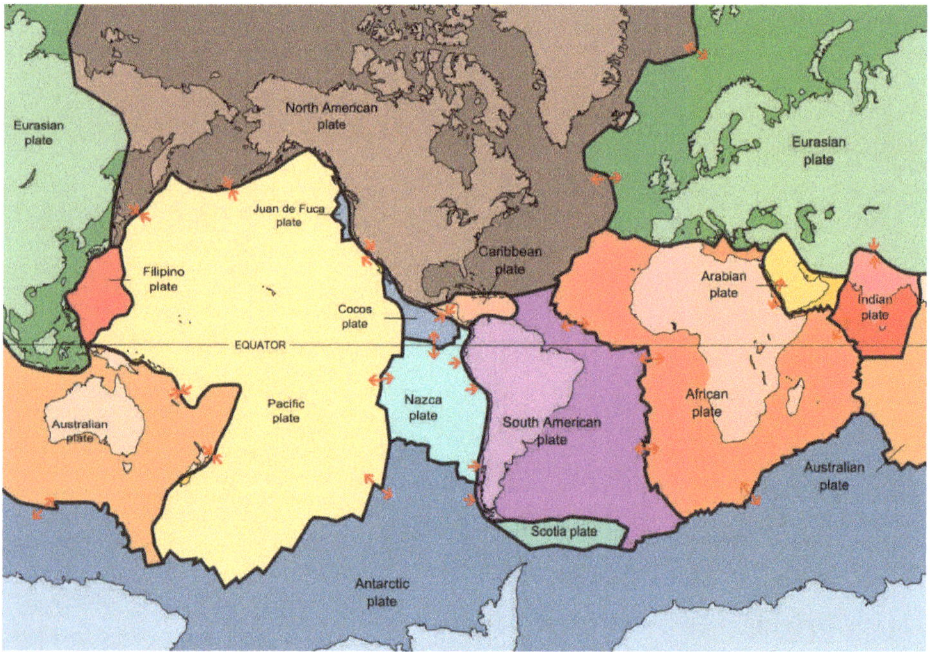

The tectonic plates of the world were mapped in the second half of the 20th century.

The lithosphere, which is the rigid outermost shell of a planet (the crust and upper mantle), is broken up into tectonic plates. The Earth's lithosphere is composed of seven or eight major plates (depending on how they are defined) and many minor plates. Where the plates meet, their relative motion determines the type of boundary: convergent, divergent, or transform. Earthquakes, volcanic activity, mountain-building, and oceanic trench formation occur along these plate boundaries. The relative movement of the plates typically ranges from zero to 100 mm annually.

Tectonic plates are composed of oceanic lithosphere and thicker continental lithosphere, each topped by its own kind of crust. Along convergent boundaries, subduction carries plates into the mantle; the material lost is roughly balanced by the formation of new (oceanic) crust along divergent margins by seafloor spreading. In this way, the total surface of the lithosphere remains the same. This prediction of plate tectonics is also referred to as the conveyor belt principle. Earlier theories (that still have some supporters) propose gradual shrinking (contraction) or gradual expansion of the globe. Tectonic plates are able to move because the Earth›s lithosphere has greater strength than the underlying asthenosphere. Lateral density variations in the mantle result in convection. Plate movement is thought to be driven by a combination of the motion of the seafloor away from the spreading ridge (due to variations in topography and density of the crust, which result in differences in gravitational forces) and drag, with downward suction, at the subduction zones. Another explanation lies in the different forces generated by tidal forces of the Sun and Moon. The relative importance of each of these factors and their relationship to each other is unclear, and still the subject of much debate.

KEY PRINCIPLES

The outer layers of the Earth are divided into the lithosphere and asthenosphere. This is based on differences in mechanical properties and in the method for the transfer of heat. Mechanically, the lithosphere is cooler and more rigid, while the asthenosphere is hotter and flows more easily. In terms of heat transfer, the lithosphere loses heat by conduction, whereas the asthenosphere also transfers heat by convection and has a nearly adiabatic temperature gradient. This division should not be confused with the *chemical* subdivision of these same layers into the mantle (comprising both the asthenosphere and the mantle portion of the lithosphere) and the crust: a given piece of mantle may be part of the lithosphere or the asthenosphere at different times depending on its temperature and pressure.

The key principle of plate tectonics is that the lithosphere exists as separate and distinct *tectonic plates*, which ride on the fluid-like (visco-elastic solid) asthenosphere. Plate motions range up to a typical 10–40 mm/year (Mid-Atlantic Ridge; about as fast as fingernails grow), to about 160 mm/year (Nazca Plate; about as fast as hair grows). The driving mechanism behind this movement is described below.

Tectonic lithosphere plates consist of lithospheric mantle overlain by either or both of two types of crustal material: oceanic crust (in older texts called *sima* from silicon and magnesium) and continental crust (*sial* from silicon and aluminium). Average oceanic lithosphere is typically 100 km (62 mi) thick; its thickness is a function of its age: as time passes, it conductively cools and subjacent cooling mantle is added to its base. Because it is formed at mid-ocean ridges and spreads outwards, its thickness is therefore a function of its distance from the mid-ocean ridge where it was formed. For a typical distance that oceanic lithosphere must travel before being subducted, the thickness varies from about 6 km (4 mi) thick at mid-ocean ridges to greater than 100 km (62 mi) at subduction zones; for shorter or longer distances, the subduction zone (and therefore also the mean) thickness becomes smaller or larger, respectively. Continental lithosphere is typically ~200 km thick, though this varies considerably between basins, mountain ranges, and stable cratonic interiors of continents. The two types of crust also differ in thickness, with continental crust being considerably thicker than oceanic (35 km vs. 6 km).

The location where two plates meet is called a *plate boundary*. Plate boundaries are commonly associated with geological events such as earthquakes and the creation of topographic features such as mountains, volcanoes, mid-ocean ridges, and oceanic trenches. The majority of the world›s active volcanoes occur along plate boundaries, with the Pacific Plate›s Ring of Fire being the most active and widely known today. These boundaries are discussed in further detail below. Some volcanoes occur in the interiors of plates, and these have been variously attributed to internal plate deformation and to mantle plumes.

As explained above, tectonic plates may include continental crust or oceanic crust, and most plates contain both. For example, the African Plate includes the continent and parts of the floor of the Atlantic and Indian Oceans. The distinction between oceanic crust and continental crust is based on their modes of formation. Oceanic crust is formed at sea-floor spreading centers, and continental crust is formed through arc volcanism and accretion of terranes through tectonic processes, though some of these terranes may contain ophiolite sequences, which are pieces of oceanic crust considered to be part of the continent when they exit the standard cycle of formation and spreading centers and subduction beneath continents. Oceanic crust is also denser than continental crust owing to their different compositions. Oceanic crust is denser because it has less silicon and more heavier elements ("mafic") than continental crust ("felsic"). As a result of this density stratification, oceanic crust generally lies below sea level (for example most of the Pacific Plate), while continental crust buoyantly projects above sea level (see the page isostasy for explanation of this principle).

TYPES OF PLATE BOUNDARIES

Three types of plate boundaries exist, with a fourth, mixed type, characterized by the way the plates move relative to each other. They are associated with different types of surface phenomena. The different types of plate boundaries are:

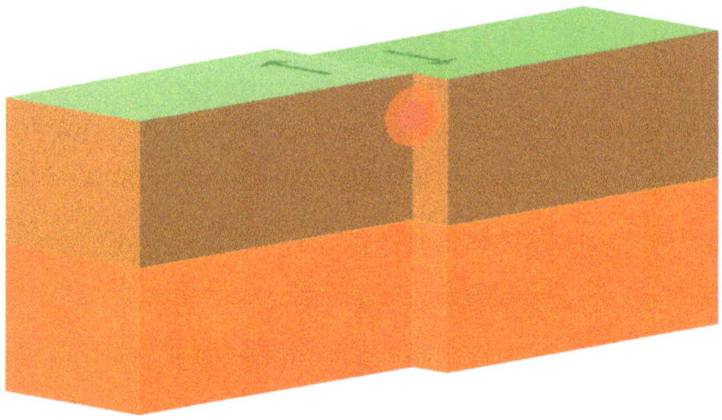

Transform boundary

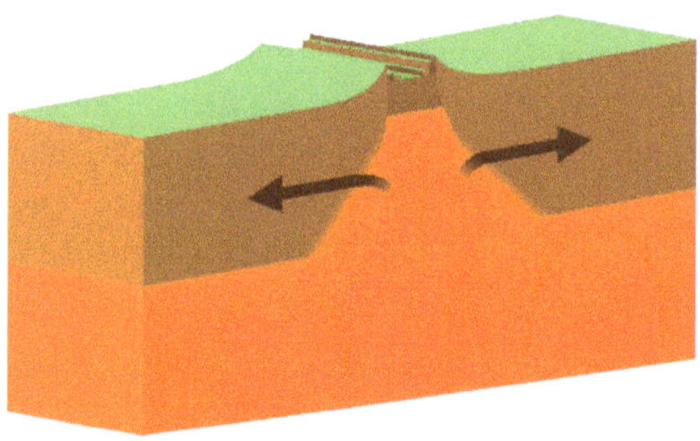

Divergent boundary

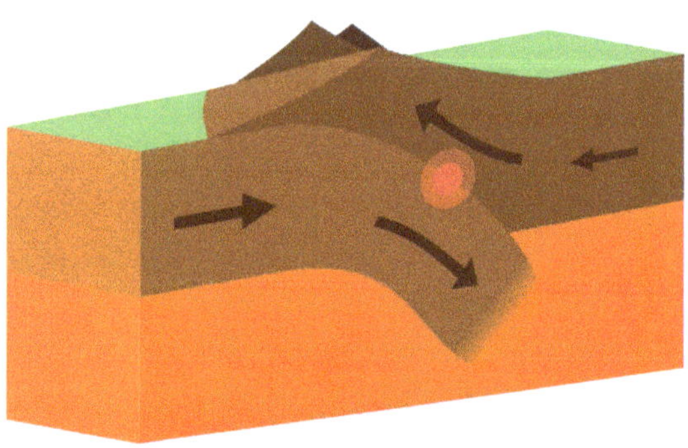

Convergent boundary.

- *Transform boundaries (Conservative)* occur where two lithospheric plates slide, or perhaps more accurately, grind past each other along transform faults, where plates are neither created nor destroyed. The relative motion of the two plates is either sinistral (left side toward the observer) or dextral (right side toward the observer). Transform faults occur across a spreading center. Strong earthquakes can occur along a fault. The San Andreas Fault in California is an example of a transform boundary exhibiting dextral motion.
- *Divergent boundaries (Constructive)* occur where two plates slide apart from each other. At zones of ocean-to-ocean rifting, divergent boundaries form by seafloor spreading, allowing for the formation of new ocean basin. As the ocean plate splits, the ridge forms at the spreading center, the ocean basin expands, and finally, the plate area increases causing many small volcanoes and/or shallow earthquakes. At zones of continent-to-continent rifting, divergent boundaries may cause new ocean basin to form as the continent splits, spreads, the central rift collapses, and ocean fills the basin. Active zones of Mid-ocean ridges (e.g., Mid-Atlantic Ridge and East Pacific Rise), and continent-to-continent rifting (such as Africa's East African Rift and Valley, Red Sea) are examples of divergent boundaries.
- *Convergent boundaries (Destructive)* (or *active margins*) occur where two plates slide toward each other to form either a subduction zone (one plate moving underneath the other) or a continental collision. At zones of ocean-to-continent subduction (e.g. the Andes mountain range in South America, and the Cascade Mountains in Western United States), the dense oceanic lithosphere plunges beneath the less dense continent. Earthquakes trace the path of the downward-moving plate as it descends into asthenosphere, a trench forms, and as the subducted plate is heated it releases volatiles, mostly water from hydrous minerals, into the surrounding mantle. The addition of water lowers the melting point of the mantle material above the subducting slab, causing it to melt. The magma that results typically leads to volcanism. At zones of ocean-to-ocean subduction (e.g. Aleutian islands, Mariana Islands, and the Japanese island arc), older, cooler, denser crust slips beneath less dense crust. This causes earthquakes and a deep trench to form in an arc shape. The upper mantle of the subducted plate then heats and magma rises to form curving chains of volcanic islands. Deep marine trenches are typically associated with subduction zones, and the basins that develop along the active boundary are often called "foreland basins". Closure of ocean basins can occur at continent-to-continent boundaries (e.g., Himalayas and Alps): collision between masses of granitic continental lithosphere; neither mass is subducted; plate edges are compressed, folded, uplifted.
- *Plate boundary zones* occur where the effects of the interactions are unclear, and the boundaries, usually occurring along a broad belt, are not well defined and may show various types of movements in different episodes.

Transform Fault

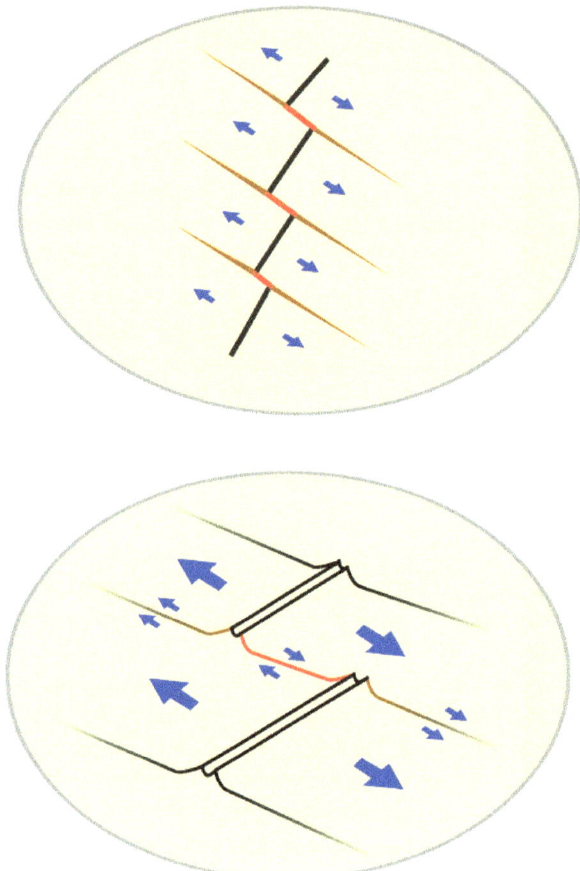

Transform fault (the red lines).

A **transform fault** or **transform boundary** (also known as a **conservative plate boundary**, since these faults neither create nor destroy lithosphere), is a type of fault whose relative motion is predominantly horizontal, in either a sinistral (left lateral) or dextral (right lateral) direction. Furthermore, transform faults end abruptly and are connected on both ends to other faults, ridges, or subduction zones. While most transform faults are hidden in the deep oceans where they offset divergent boundaries as series of short zigzags accommodating seafloor spreading, the best-known (and most destructive) are those on land at the margins of tectonic plates. Transform faults are the only type of strike-slip fault that can be classified as a plate boundary.

John Tuzo Wilson recognized that the offsets of oceanic ridges by faults do not follow the classical pattern of an offset fence or geological marker in Reid'srebound theory of faulting, from which the sense of slip is derived. The new class of faults, called transform faults, produce slip in the opposite direction from what one would surmise from the standard interpretation of an offset geological feature. Slip along transform faults does not increase the distance between the ridges it separates; the distance remains constant in earthquakes because the ridges are spreading centers. This hypothesis was confirmed in a study of the fault plane solutions that showed the slip on transform faults points in the opposite direction than classical interpretation would suggest.

Difference Between Transform and Transcurrent Faults

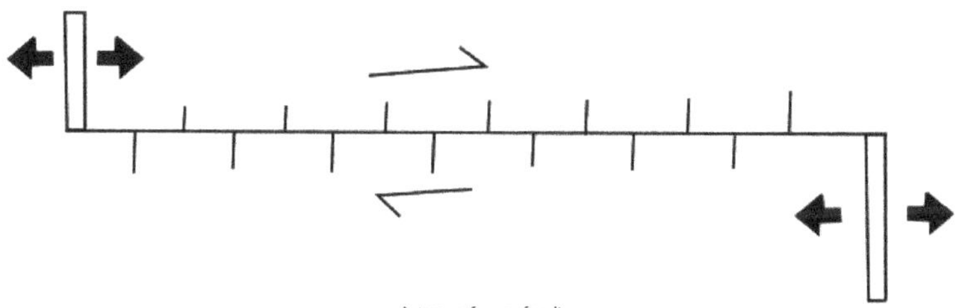

A transform fault
Linked on both ends by another tectonic boundary structure. As well as having equal deformation along the entire length of the fault line.

Transform fault.

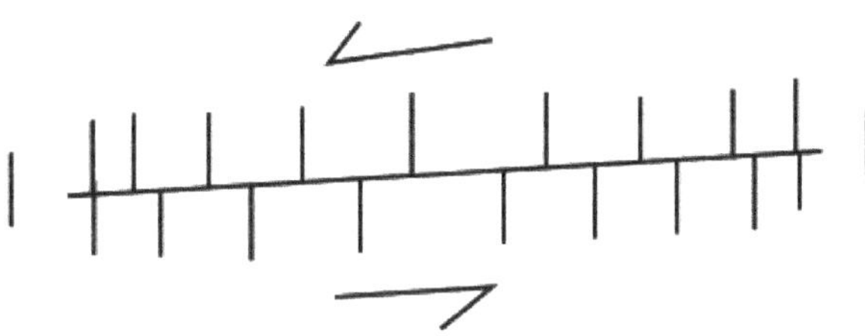

A transcurrent fault with unequal deformation.

Transcurrent fault.

Transform faults are closely related to transcurrent faults, and are commonly confused. Both types of faults are strike-slip or side-to-side in movement; however, transform faults end at the junction of another plate boundary or fault type, while transcurrent faults die out without a junction. In addition, transform faults have equal deformation across the entire fault line, while transcurrent faults have greater displacement in the middle of the fault zone and less on the margins. Finally, transform faults can form a tectonic plate boundary, while transcurrent faults cannot.

Mechanics

The effect of a fault is to relieve strain, which can be caused by compression, extension, or lateral stress in the rock layers at the surface or deep in the Earth's subsurface. Transform faults specifically relieve strain by transporting the strain between ridges or subduction zones. Transform faults also act as the plane of weakness allowing for the splitting in rift zones.

Examples

Transform faults are commonly found linking segments of mid-oceanic ridges or spreading centres. These mid-oceanic ridges are where new seafloor is constantly created through the upwelling of new basaltic magma. With new seafloor being pushed and pulled out, the older seafloor slowly slides away from the mid-oceanic ridges toward the continents. Although separated only by tens of kilometers, this separation between segments of the ridges causes portions of the seafloor to push past each other in opposing directions. This lateral movement of seafloors past each other is where transform faults are currently active.

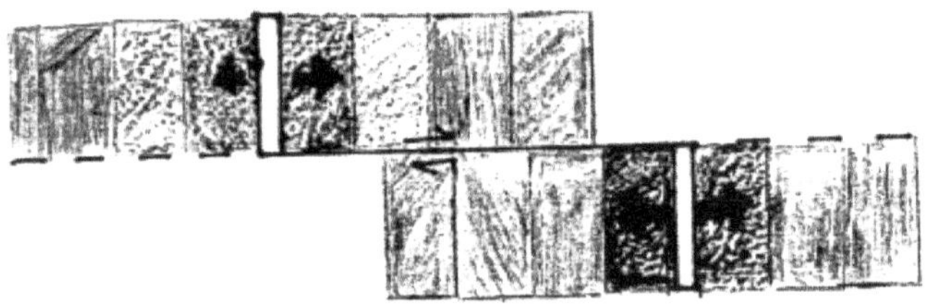

The blocks represent different magnetic strips in the sea floor moving in opposite directions to each other. The solid line represents the active transform fault and the dashed lines represent the fracture zones (inactive areas).

Spreading center and strips.

Transform faults move differently than a strike-slip fault at the mid-oceanic ridge. Instead of the ridges moving away from each other, like other strike-slip faults, transform fault ridges will stay in the same fixed location, and the new ocean seafloor being created at the ridges is pushed away from the ridge. Evidence of this can be found in paleomagnetic striping on the seafloor.

A paper written by Gerya theorizes that the creation of the transform faults between the ridges of the mid-oceanic ridge is attributed to rotated and stretched sections of the mid-oceanic ridge. This occurs over a long period of time with the spreading center or ridge slowly deforming from a straight line to a curved line. Finally, fracturing along these planes forms transform faults. As this takes place, the fault changes from a normal fault with extensional stress to a strike slip fault with lateral stress. In the study done by Bonatti & Crane, peridotite and gabbro rocks were discovered in the edges of the transform ridges. These rocks are created deep inside the Earth's mantle and then rapidly exhumed to the surface. This evidence helps to prove that new seafloor is being created at the mid-oceanic ridges and further supports the theory of plate tectonics.

As previously stated, active transform faults are between two tectonic structures or faults. Fracture zones represent the previously active transform fault lines, which have since passed the active transform zone and are being pushed toward the continents. These elevated ridges on the ocean floor can be traced for hundreds of miles and in some cases even from one continent across an ocean to the other continent.

The most prominent examples of the mid-oceanic ridge transform zones are located in Atlantic Ocean between South America and Africa. Known as the St. Paul, Romanche, Chain, and Ascension fracture zones, these areas have with deep, easily identifiable transform faults and ridges. Other locations include: the East Pacific Ridge located in the South Eastern Pacific Ocean, which meets up with San Andreas Fault to the North.

Transform faults are not limited to oceanic crust and spreading centers; many transform faults are located on continental margins. The best example is the San Andreas Fault on the Pacific coast of the United States. The San Andreas Fault links the East Pacific Rise off of the West coast of Mexico (Gulf of California) to the Mendocino Triple Junction (Part of the Juan de Fuca plate) located off the coast of the North Western United States making it a ridge-to-transform style transform fault. The formation of the San Andreas Fault system occurred fairly recently during the Oligocene Period between 34 million and 24 million years ago. During this period, the Farallon plate, followed by the Pacific plate, collided into the North American plate. The collision led to the subduction of the Farallon plate underneath the North American plate. Once the spreading center separating the Pacific and Farallon plate was subducted underneath the North American plate, the San Andreas Continental Transform Fault system was created.

The Southern Alps rise dramatically beside the Alpine Fault on New Zealand›s West Coast. About 500 kilometres (300 mi) long; northwest at top.

Other examples include:

- Middle East›s Dead Sea Transform fault
- New Zealand›s Alpine Fault
- Pakistan›s Chaman Fault
- Turkey›s North Anatolian Fault
- North America›s Queen Charlotte Fault

Transform fault types

In his groundbreaking work on transform fault systems, Tuzo Wilson said that transform faults must be connected to other faults or tectonic plate boundaries on both ends; because of that requirement, transform faults can grow in length, keep a constant length, or decrease in length. These length changes are dependent on which type of faults or tectonic structures connect with the transform fault. With this in mind, Wilson described six types of transform faults: Growing length faults: In situations where a transform fault links together a spreading center and the upper block of a subduction zone or when two upper blocks of subduction zones are linked the transform fault itself will grow in length.

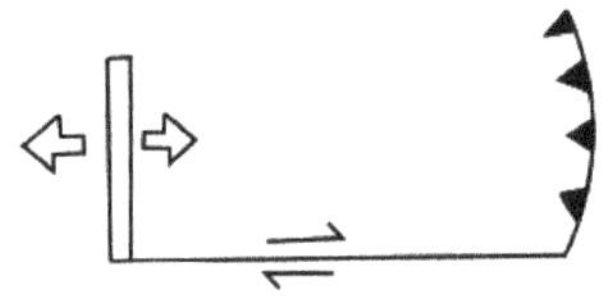

A spreading center (left) connected by a transform fault to the upper slab of a subduction zone

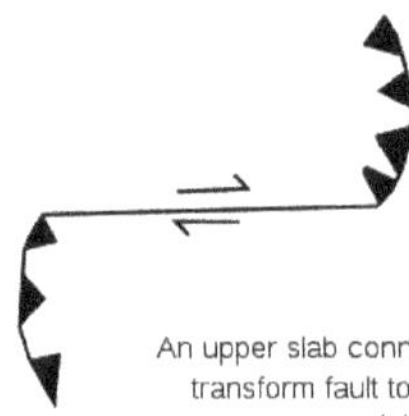

An upper slab connected by a transform fault to another upper slab.

Constant length faults: In other cases, transform faults will remain at a constant length. This consistency can be attributed to many different reasons. In the case of a ridge-to-ridge transforms, it is caused by the continuous growth by both ridges outward, canceling any change in length. The opposite occurs when a ridge linked to a subducting plate, where all the lithosphere (new sea floor) being created by the ridge is being subducted, or swallowed up, by the subduction zone. Finally, when two upper subduction plates are linked there is no change in length. This is due to the plates moving parallel with each other and no new lithosphere is being created to change that length.

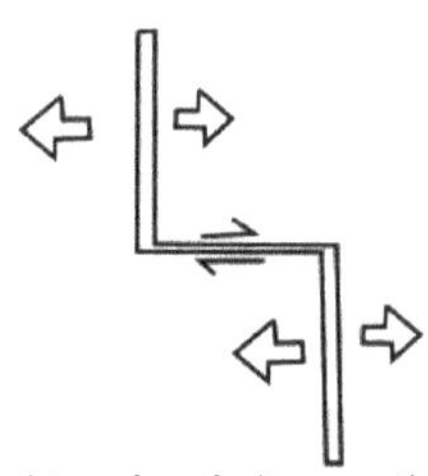

A transform fault connecting two mid-oceanic ridges.

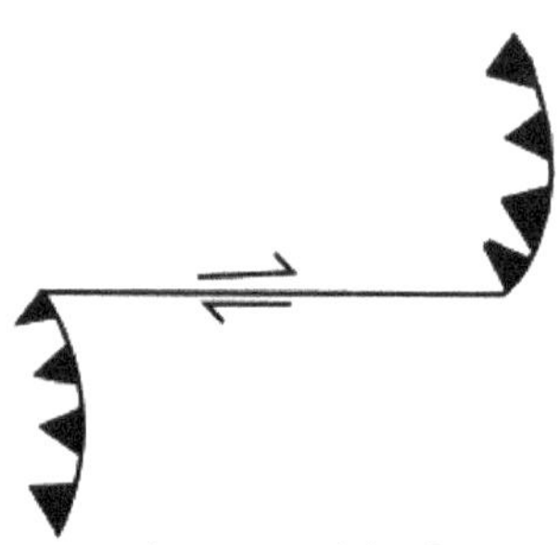

An upper slab of a subduction connected by a transform fault to decending slab of a subduction zone

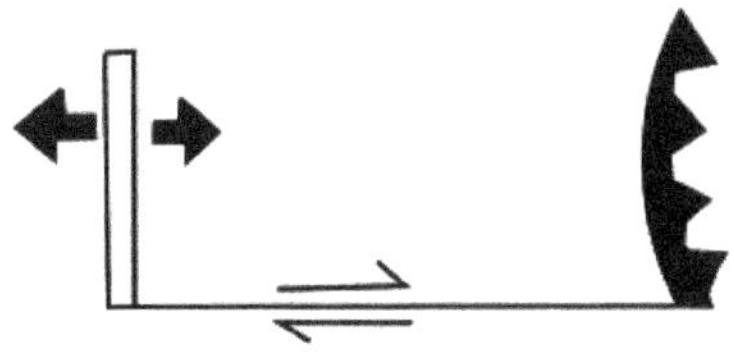

A spreading center (left) connected by a transform fault to the descending slab of a subduction zone.

Decreasing length faults: In rare cases, transform faults can shrink in length. These occur when two descending subduction plates are linked by a transform fault. In time as the plates are subducted, the transform fault will decrease in length until the transform fault disappears completely, leaving only two subduction zones facing in opposite directions.

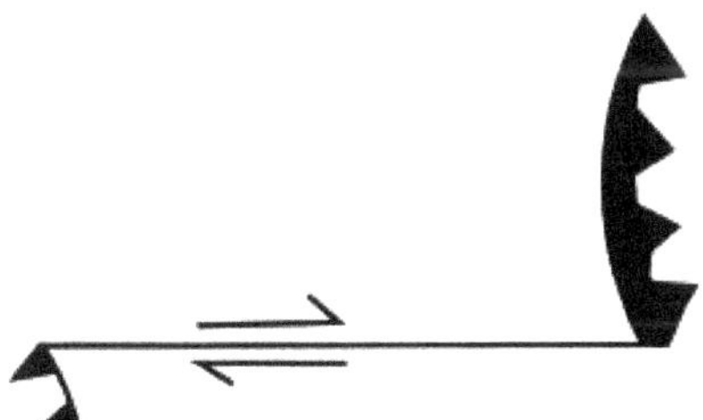

A descending slab connected by a transform fault to another descending slab.

Divergent Boundary

In plate tectonics, a **divergent boundary** or **divergent plate boundary** (also known as a **constructive boundary** or an **extensional boundary**) is a linear feature that exists between two tectonic plates that are moving away from each other. Divergent boundaries within continents initially produce rifts which eventually become rift valleys. Most active divergent plate boundaries occur between oceanic plates and exist as mid-oceanic ridges. Divergent boundaries also form volcanic islands which occur when the plates move apart to produce gaps which molten lava rises to fill.

Current research indicates that complex convection within the Earth›s mantle allows material to rise to the base of the lithosphere beneath each divergent plate boundary. This supplies the area with vast amounts of heat and a reduction in pressure that melts rock from the asthenosphere (or upper mantle) beneath the rift area forming large flood basalt or lava flows. Each eruption occurs in only a part of the plate boundary at any one time, but when it does occur, it fills in the opening gap as the two opposing plates move away from each other. Over millions of years, tectonic plates may move many hundreds of kilometers away from both sides of a divergent plate boundary. Because of this, rocks closest to a boundary are younger than rocks further away on the same plate.

Description

Bridge across the Álfagjá rift valley in southwest Iceland, that is part of the boundary between the Eurasian and North American continental tectonic plates.

At divergent boundaries, two plates move apart from each other and the space that this creates is filled with new crustal material sourced from molten magma that forms below. The origin of new divergent boundaries at triple junctions is sometimes thought to be associated with the phenomenon known as hotspots. Here, exceedingly large convective cells bring very large quantities of hot asthenospheric material near the surface and the kinetic energy is thought to be sufficient to break apart the lithosphere. The hot spot which may have initiated the Mid-Atlantic Ridge system currently underlies Iceland which is widening at a rate of a few centimeters per year.

Divergent boundaries are typified in the oceanic lithosphere by the rifts of the oceanic ridge system, including the Mid-Atlantic Ridge and the East Pacific Rise, and in the continental lithosphere by rift valleys such as the famous East African Great Rift Valley. Divergent boundaries can create massive fault zones in the oceanic ridge system. Spreading is generally not uniform, so where spreading rates of adjacent ridge blocks are different, massive transform faults occur. These are the fracture zones, many bearing names, that are a major source of submarine earthquakes. A sea floor map will show a rather strange pattern of blocky structures that are separated by linear features perpendicular to the ridge axis. If one views the sea floor between the fracture zones as conveyor belts carrying the ridge on each side of the rift away from the spreading center the action becomes clear. Crest depths of the old ridges, parallel to the current spreading center, will be older and deeper... (from thermal contraction and subsidence).

It is at mid-ocean ridges that one of the key pieces of evidence forcing acceptance of the seafloor spreading hypothesis was found. Airborne geomagnetic surveys showed a strange pattern of symmetrical magnetic reversals on opposite sides of ridge centers. The pattern was far too regular to be coincidental as the widths of the opposing bands were too closely matched. Scientists had been studying polar reversals and the link was made by Lawrence W. Morley, Frederick John Vine and Drummond Hoyle Matthews in the Morley–Vine–Matthews hypothesis. The magnetic banding directly corresponds with the Earth›s polar reversals. This was confirmed by measuring the ages of the rocks within each band. The banding furnishes a map in time and space of both spreading rate and polar reversals.

Convergent Boundary

In plate tectonics, a **convergent boundary**, also known as a **destructive plate boundary** (because of subduction), is an actively deforming region where two (or more) tectonic plates or fragments of the lithosphere move toward one another and collide. As a result of pressure, friction, and plate material melting in the mantle, earthquakes and volcanoes are common near convergent boundaries. When two plates move towards one another, they form either a subduction zone or a continental collision. This depends on the nature of the plates involved. In a subduction zone, the subducting plate, which is normally a plate with oceanic crust, moves beneath the other plate, which can be made of either oceanic or continental crust. During collisions between two continental plates, large mountain ranges, such as the Himalayas are formed.

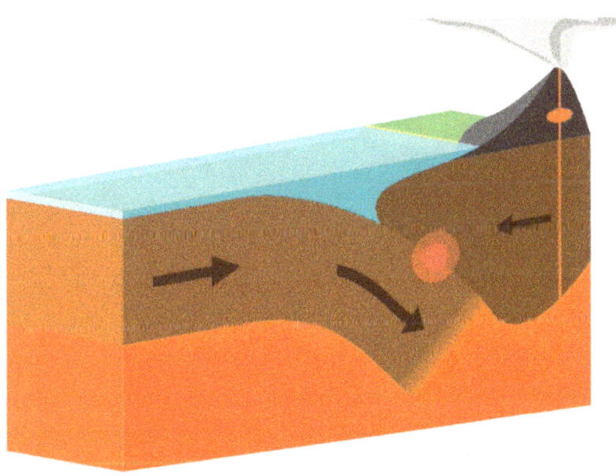

Oceanic-continental

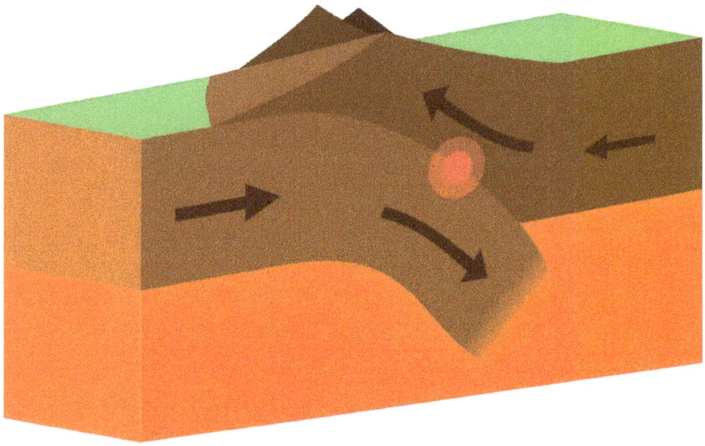

Continental-continental.

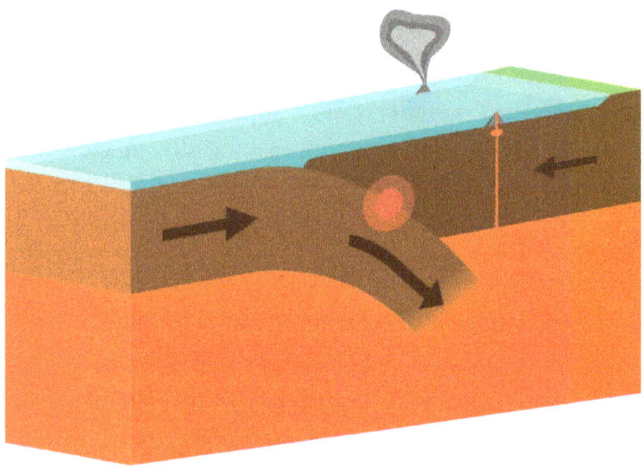

Oceanic-oceanic.

Descriptions

The nature of a convergent boundary depends on the type of plates that are colliding. At an oceanic-continental convergent boundary, the oceanic lithosphere will always subduct below the continental lithosphere. This is caused by the relative density difference between the oceanic (3.0 g/cm3) and continental (2.7 g/cm3) lithosphere. This type of boundary is also called a subduction zone. At the surface, the topographic expression is commonly an oceanic trench which forms on the oceanic side, parallel to the subduction zone. On the continental side, a chain of volcanoes forms above the location of the subducting plate, creating a mountain chain referred to as a volcanic arc. An example of a continental-oceanic subduction zone is the area along the western coast of South America where the oceanic Nazca Plate is being subducted beneath the continental South American Plate. A volcanic arc is formed on the continental plate, above the location of the downgoing oceanic slab. The volcanic arc is the surface expression of the magma that is generated by hydrous melting of the mantle above the downgoing slab. Hydrated minerals (e.g., phlogopite, lawsonite, amphibole) within the oceanic lithosphere become unstable at certain depths due to increased temperature and pressure, causing the crystal structure of the hydrated minerals break down and release water. The buoyant fluids then rise into the asthenosphere, where they lower the melting temperature of the mantle and cause partial melting. These melts rise to the surface and are the source of some of the most explosive volcanism on Earth because of their high volumes of extremely pressurized aeither buckle and compress or (in some cases) one plate delves called subduction, under the other. Either action will create extensive mountain ranges. The most dramatic effect seen is where the northern margin of the Indian Plate is being thrust under a portion of the Eurasian Plate, lifting it and creating the Himalayas and the Tibetan Plateau beyond. It may have also pushed nearby parts of the Asian continent aside to the east.

When two plates with oceanic crust converge, they typically create an island arc as one plate is subducted below the other. The arc is formed from volcanoes which erupt through the overriding plate as the descending plate melts below it. The arc shape occurs because of the spherical surface of the earth (nick the peel of an orange with a knife and note the arc formed by the straight-edge of the knife). A deep oceanic trench is located in front of such arcs where the descending slab dips downward, such as the Mariana trench near the Mariana Islands.

Plates may collide at an oblique angle rather than head-on to each other (e.g. one plate moving north, the other moving south-east), and this may cause strike-slip faulting along the collision zone, in addition to subduction or compression.

Not all plate boundaries are easily defined. Some are broad belts whose movements are unclear to scientists. One example is the Mediterranean-Alpine boundary, which involves two major plates (African and Eurasian) and several micro plates. The boundaries of the plates do not necessarily coincide with those of the continents. For instance, the North American Plate covers not only North America, but also far northeastern Siberia, plus a substantial portion of the Atlantic Ocean.

Convergent Margins

A subduction zone is formed at a convergent plate boundary when one or both of the tectonic plates is composed of oceanic crust. The denser plate, made of oceanic crust, is subducted underneath the less dense plate, which can be either continental or oceanic crust. When both of the plates are made of oceanic crust, convergence is associated with island arcs such as the Solomon Islands.

An oceanic trench is found where the denser plate is subducted underneath the other plate. There is water in the rocks of the oceanic plate (because they are underwater), and as this plate moves further down into the subduction zone, much of the water contained in the plate is squeezed out when the plate begins to subduct. However, the recrystallization of ocean floor rocks, such as serpentinite, which are unstable in the upper mantle, recrystallize into olivine, causing dehydration through loss of hydroxyl groups. This addition of water to the mantle causes partial melting of the mantle, generating magma, which then rises, and which normally results in volcanoes. This normally happens at a certain depth, about 70 to 80 miles below the Earth›s surface, and so volcanoes are formed fairly close to, but not right next to, the trench.

Some convergent margins have zones of active seafloor spreading behind the island arc, known as back-arc basins. When one plate is composed of oceanic lithosphere and the other is composed of continental lithosphere, the denser oceanic plate is subducted, often forming an orogenic belt and associated mountain range. This type of convergent boundary is similar to the Andes or the Cascade Range in North America.

When two plates containing continental crust collide, both are too light to subduct. In this case, a continent-continent collision occurs, creating especially large mountain ranges. The most spectacular example of this is the Himalayas.

When the subducting plate approaches the trench obliquely, the convergent plate boundary includes a major component of strike-slip faulting within the over-riding plate. The best example of this is the Sumatra convergent margin, where orthogonal convergence on the Sunda megathrust is occurring intermixed with movement on the Great Sumatran fault.

DRIVING FORCES OF PLATE MOTION

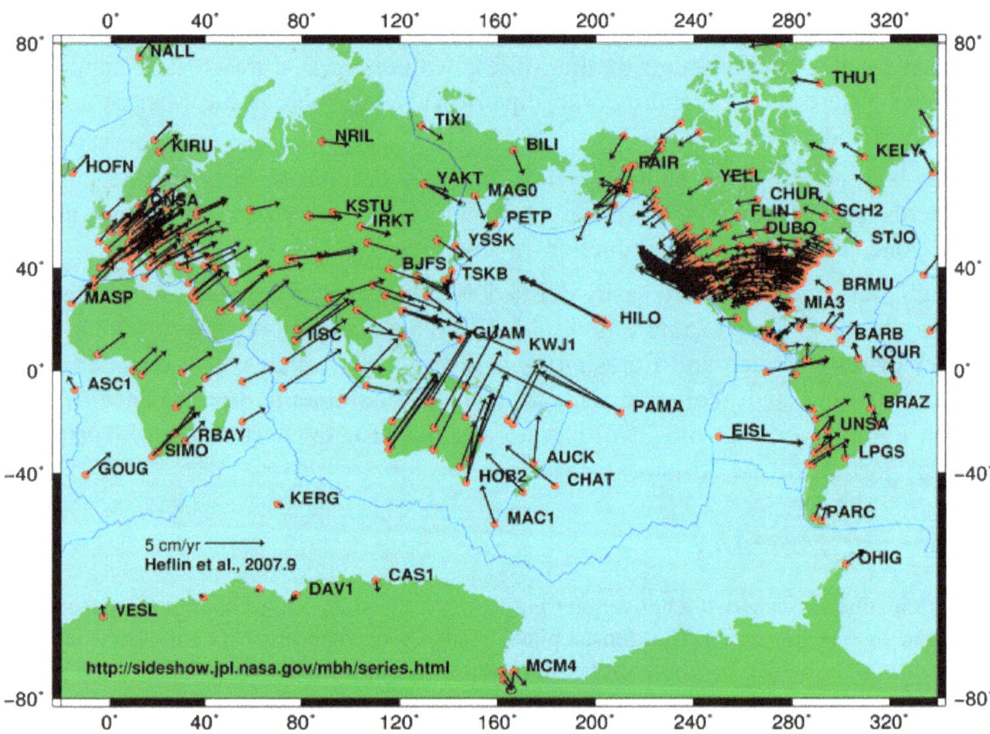

Plate motion based on Global Positioning System (GPS) satellite data from NASA JPL. The vectors show direction and magnitude of motion.

It has generally been accepted that tectonic plates are able to move because of the relative density of oceanic lithosphere and the relative weakness of the asthenosphere. Dissipation of heat from the mantle is acknowledged to be the original source of the energy required to drive plate tectonics through convection or large scale upwelling and doming. The current view, though still a matter of some debate, asserts that as a consequence, a powerful source of plate motion is generated due to the excess density of the oceanic lithosphere sinking in subduction zones. When the new crust forms at mid-ocean ridges, this oceanic lithosphere is initially less dense than the underlying asthenosphere, but it becomes denser with age as it conductively cools and thickens. The greater density of old lithosphere relative to the underlying asthenosphere allows it to sink into the deep mantle at subduction zones, providing most of the driving force for plate movement. The weakness of the asthenosphere allows the tectonic plates to move easily towards a subduction zone. Although subduction is thought to be the strongest force driving plate motions, it cannot be the only force since there are plates such as the North American Plate which are moving, yet are nowhere being subducted. The same is true for the enormous Eurasian Plate. The sources of plate motion are a matter of intensive research and discussion among scientists. One of the main points is that the kinematic pattern of the movement itself should be separated clearly from the possible geodynamic mechanism that is invoked as the driving force of the observed movement, as some patterns may be explained by more than one mechanism. In short, the driving forces advocated at the moment can be divided into three categories based on the relationship to the movement: mantle dynamics related, gravity related (mostly secondary forces).

Driving Forces Related to Mantle Dynamics

For much of the last quarter century, the leading theory of the driving force behind tectonic plate motions envisaged large scale convection currents in the upper mantle, which can be transmitted through the asthenosphere. This theory was launched by Arthur Holmes and some forerunners in the 1930s and was immediately recognized as the solution for the acceptance of the theory as originally discussed in the papers of Alfred Wegener in the early years of the century. However, despite its acceptance, it was long debated in the scientific community because the leading ("fixist") theory still envisaged a static Earth without moving continents up until the major breakthroughs of the early sixties.

Two- and three-dimensional imaging of Earth›s interior (seismic tomography) shows a varying lateral density distribution throughout the mantle. Such density variations can be material (from rock chemistry), mineral (from variations in mineral structures), or thermal (through thermal expansion and contraction from heat energy). The manifestation of this varying lateral density is mantle convection from buoyancy forces.

How mantle convection directly and indirectly relates to plate motion is a matter of ongoing study and discussion in geodynamics. Somehow, this energy must be transferred to the lithosphere for tectonic plates to move. There are essentially two main types of forces that are thought to influence plate motion: friction and gravity.

- Basal drag (friction): Plate motion driven by friction between the convection currents in the asthenosphere and the more rigid overlying lithosphere.
- Slab suction (gravity): Plate motion driven by local convection currents that exert a downward pull on plates in subduction zones at ocean trenches. Slab suction may occur in a geodynamic setting where basal tractions continue to act on the plate as it dives into the mantle (although perhaps to a greater extent acting on both the under and upper side of the slab).

Lately, the convection theory has been much debated as modern techniques based on 3D seismic tomography still fail to recognize these predicted large scale convection cells. Therefore, alternative views have been proposed: In the theory of plume tectonics developed during the 1990s, a modified concept of mantle convection currents is used. It asserts that super plumes rise from the deeper mantle and are the drivers or substitutes of the major convection cells. These ideas, which find their roots in the early 1930s with the so-called "fixistic" ideas of the European and Russian Earth Science Schools, find resonance in the modern theories which envisage hot spots/mantle plumes which remain fixed and are overridden by oceanic and continental lithosphere plates over time and leave their traces in the geological record (though these phenomena are not invoked as real driving mechanisms, but rather as modulators). Modern theories that continue building on the older mantle doming concepts and see plate movements as a secondary phenomena are beyond the scope of this page and are discussed elsewhere (for example on the plume tectonics page).

Another theory is that the mantle flows neither in cells nor large plumes but rather as a series of channels just below the Earth›s crust, which then provide basal friction to the lithosphere. This theory, called "surge tectonics", became quite popular in geophysics and geodynamics during the 1980s and 1990s. Recent research, based on three-dimensional computer modeling, suggests that plate geometry is governed by a feedback between mantle convection patterns and the strength of the lithosphere.

Driving Forces Related to Gravity

Forces related to gravity are usually invoked as secondary phenomena within the framework of a more general driving mechanism such as the various forms of mantle dynamics described above.

Gravitational sliding away from a spreading ridge: According to many authors, plate motion is driven by the higher elevation of plates at ocean ridges. As oceanic lithosphere is formed at spreading ridges from hot mantle material, it gradually cools and thickens with age (and thus adds distance from the ridge). Cool oceanic lithosphere is significantly denser than the hot mantle material from which it is derived and so with increasing thickness it gradually subsides into the mantle to compensate the greater load. The result is a slight lateral incline with increased distance from the ridge axis.

This force is regarded as a secondary force and is often referred to as "ridge push". This is a misnomer as nothing is "pushing" horizontally and tensional features are dominant along ridges. It is more accurate to refer to this mechanism as gravitational sliding as variable topography across the totality of the plate can vary considerably and the topography of spreading ridges is only the most prominent feature. Other mechanisms generating this gravitational secondary force include flexural bulging of the lithosphere before it dives underneath an adjacent plate which produces a clear topographical feature that can offset, or at least affect, the influence of topographical ocean ridges, and mantle plumes and hot spots, which are postulated to impinge on the underside of tectonic plates.

Slab-pull: Current scientific opinion is that the asthenosphere is insufficiently competent or rigid to directly cause motion by friction along the base of the lithosphere. Slab pull is therefore most widely thought to be the greatest force acting on the plates. In this current understanding, plate motion is mostly driven by the weight of cold, dense plates sinking into the mantle at trenches. Recent models indicate that trench suction plays an important role as well. However, as the North American Plate is nowhere being subducted, yet it is in motion presents a problem. The same holds for the African, Eurasian, and Antarctic plates. Gravitational sliding away from mantle doming: According to older theories, one of the driving mechanisms of the plates is the existence of large scale asthenosphere/ mantle domes which cause the gravitational sliding of lithosphere plates away from them. This gravitational sliding represents a secondary phenomenon of this basically vertically oriented mechanism. This can act on various scales, from the small scale of one island arc up to the larger scale of an entire ocean basin.

Driving Forces related to Earth Rotation

Alfred Wegener, being a meteorologist, had proposed tidal forces and pole flight force as the main driving mechanisms behind continental drift; however, these forces were considered far too small to cause continental motion as the concept then was of continents plowing through oceanic crust. Therefore, Wegener later changed his position and asserted that convection currents are the main driving force of plate tectonics in the last edition of his book in 1929.

However, in the plate tectonics context (accepted since the seafloor spreading proposals of Heezen, Hess, Dietz, Morley, Vine, and Matthews (see below) during the early 1960s), the oceanic crust is suggested to be in motion *with* the continents which caused the proposals related to Earth rotation to be reconsidered. In more recent literature, these driving forces are:

- Tidal drag due to the gravitational force the Moon (and the Sun) exerts on the crust of the Earth
- Global deformation of the geoid due to small displacements of rotational pole with respect to the Earth's crust;
- Other smaller deformation effects of the crust due to wobbles and spin movements of the Earth rotation on a smaller time scale.

Forces that are small and generally negligible are:

- The Coriolis force
- The centrifugal force, which is treated as a slight modification of gravity[249]

For these mechanisms to be overall valid, systematic relationships should exist all over the globe between the orientation and kinematics of deformation and the geographical latitudinal and longitudinal grid of the Earth itself. Ironically, these systematic relations studies in the second half of the nineteenth century and the first half of the twentieth century underline exactly the opposite: that the plates had not moved in time, that the deformation grid was fixed with respect to the Earth equator and axis, and that gravitational driving forces were generally acting vertically and caused only local horizontal movements (the so-called pre-plate tectonic, "fixist theories"). Later studies (discussed below on this page), therefore, invoked many of the relationships recognized during this pre-plate tectonics period to support their theories (see the anticipations and reviews in the work of van Dijk and collaborators).

Of the many forces discussed in this paragraph, tidal force is still highly debated and defended as a possible principle driving force of plate tectonics. The other forces are only used in global geodynamic models not using plate tectonics concepts (therefore beyond the discussions treated in this section) or proposed as minor modulations within the overall plate tectonics model. In 1973, George W. Moore of the USGS and R. C. Bostrom presented evidence for a general westward drift of the Earth's lithosphere with respect to the mantle. He concluded that tidal forces (the tidal lag or "friction") caused by the Earth's rotation and the forces acting upon it by the Moon are a driving force for plate tectonics. As the Earth spins eastward beneath the moon, the moon's gravity ever so slightly pulls the Earth's surface layer back westward, just as proposed by Alfred Wegener (see above). In a more recent 2006 study, scientists reviewed and advocated these earlier proposed ideas. It has also been suggested recently in Lovett (2006) that this observation may also explain why Venus and Mars have no plate tectonics, as Venus has no moon and Mars' moons are too small to have significant tidal effects on the planet. In a recent paper, it was suggested that, on the other hand, it can easily be observed that many plates are moving north and eastward, and that the dominantly westward motion of the Pacific Ocean basins derives simply from the eastward bias of the Pacific spreading center (which is not a predicted manifestation of such lunar forces). In the same paper the authors admit, however, that relative to the lower mantle, there is a slight westward component in the motions of all the plates. They demonstrated though that the westward drift, seen only for the past 30 Ma, is attributed to the increased dominance of the steadily growing and accelerating Pacific plate. The debate is still open.

Relative Significance of each Driving Force Mechanism

The vector of a plate's motion is a function of all the forces acting on the plate; however, therein lies the problem regarding the degree to which each process contributes to the overall motion of each tectonic plate.

The diversity of geodynamic settings and the properties of each plate result from the impact of the various processes actively driving each individual plate. One method of dealing with this problem is to consider the relative rate at which each plate is moving as well as the evidence related to the significance of each process to the overall driving force on the plate.

One of the most significant correlations discovered to date is that lithospheric plates attached to downgoing (subducting) plates move much faster than plates not attached to subducting plates. The Pacific plate, for instance, is essentially surrounded by zones of subduction (the so-called Ring of Fire) and moves much faster than the plates of the Atlantic basin, which are attached (perhaps one could say ‹welded›) to adjacent continents instead of subducting plates. It is thus thought that forces associated with the downgoing plate (slab pull and slab suction) are the driving forces which determine the motion of plates, except for those plates which are not being subducted. The driving forces of plate motion continue to be active subjects of on-going research within geophysics and tectonophysics.

DEVELOPMENT OF THE THEORY

Summary

In line with other previous and contemporaneous proposals, in 1912 the meteorologist Alfred Wegener amply described what he called continental drift, expanded in his 1915 book *The Origin of Continents and Oceans* and the scientific debate started that would end up fifty years later in the theory of plate tectonics. Starting from the idea (also expressed by his forerunners) that the present continents once formed a single land mass (which was called Pangea later on) that drifted apart, thus releasing the continents from the Earth's mantle and likening them to "icebergs" of low density granite floating on a sea of denser basalt. Supporting evidence for the idea came from the dove-tailing outlines of South America›s east coast and Africa›s west coast, and from the matching of the rock formations along these edges. Confirmation of their previous contiguous nature also came from the fossil plants *Glossopteris* and *Gangamopteris*, and the therapsid or mammal-like reptile *Lystrosaurus*, all widely distributed over South America, Africa, Antarctica, India and Australia. The evidence for such an erstwhile joining of these continents was patent to field geologists working in the southern hemisphere. The South African Alex du Toit put together a mass of such information in his 1937 publication *Our Wandering Continents*, and went further than Wegener in recognising the strong links between the Gondwana fragments.

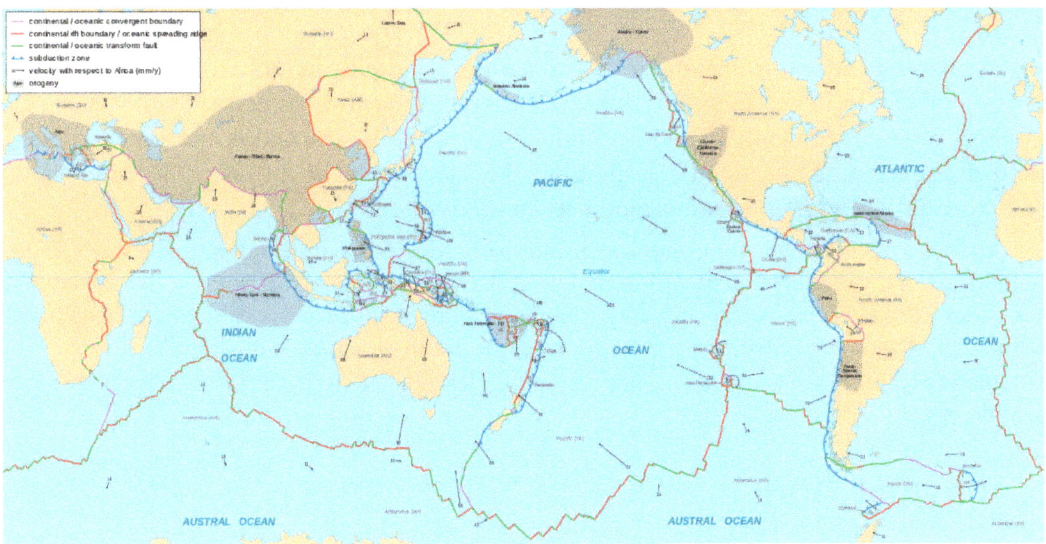

Detailed map showing the tectonic plates with their movement vectors.

But without detailed evidence and a force sufficient to drive the movement, the theory was not generally accepted: the Earth might have a solid crust and mantle and a liquid core, but there seemed to be no way that portions of the crust could move around. Distinguished scientists, such as Harold Jeffreys and Charles Schuchert, were outspoken critics of continental drift.

Despite much opposition, the view of continental drift gained support and a lively debate started between "drifters" or "mobilists" (proponents of the theory) and "fixists" (opponents). During the 1920s, 1930s and 1940s, the former reached important milestones proposing that convection currents might have driven the plate movements, and that spreading may have occurred below the sea within the oceanic crust. Concepts close to the elements now incorporated in plate tectonics were proposed by geophysicists and geologists (both fixists and mobilists) like Vening-Meinesz, Holmes, and Umbgrove.

One of the first pieces of geophysical evidence that was used to support the movement of lithospheric plates came from paleomagnetism. This is based on the fact that rocks of different ages show a variable magnetic field direction, evidenced by studies since the mid–nineteenth century. The magnetic north and south poles reverse through time, and, especially important in paleotectonic studies, the relative position of the magnetic north pole varies through time. Initially, during the first half of the twentieth century, the latter phenomenon was explained by introducing what was called "polar wander" (see apparent polar wander), i.e., it was assumed that the north pole location had been shifting through time. An alternative explanation, though, was that the continents had moved (shifted and rotated) relative to the north pole, and each continent, in fact, shows its own "polar wander path". During the late 1950s it was successfully shown on two occasions that these data could show the validity of continental drift: by Keith Runcorn in a paper in 1956, and by Warren Carey in a symposium held in March 1956.

The second piece of evidence in support of continental drift came during the late 1950s and early 60s from data on the bathymetry of the deep ocean floors and the nature of the oceanic crust such as magnetic properties and, more generally, with the development of marine geology which gave evidence for the association of seafloor spreading along the mid-oceanic ridges and magnetic field reversals, published between 1959 and 1963 by Heezen, Dietz, Hess, Mason, Vine & Matthews, and Morley.

Simultaneous advances in early seismic imaging techniques in and around Wadati-Benioff zones along the trenches bounding many continental margins, together with many other geophysical (e.g. gravimetric) and geological observations, showed how the oceanic crust could disappear into the mantle, providing the mechanism to balance the extension of the ocean basins with shortening along its margins.

All this evidence, both from the ocean floor and from the continental margins, made it clear around 1965 that continental drift was feasible and the theory of plate tectonics, which was defined in a series of papers between 1965 and 1967, was born, with all its extraordinary explanatory and predictive power. The theory revolutionized the Earth sciences, explaining a diverse range of geological phenomena and their implications in other studies such as paleogeography and paleobiology.

Continental Drift

In the late 19th and early 20th centuries, geologists assumed that the Earth's major features were fixed, and that most geologic features such as basin development and mountain ranges could be explained by vertical crustal movement, described in what is called the geosynclinal theory. Generally, this was placed in the context of a contracting planet Earth due to heat loss in the course of a relatively short geological time.

It was observed as early as 1596 that the opposite coasts of the Atlantic Ocean—or, more precisely, the edges of the continental shelves—have similar shapes and seem to have once fitted together.

Since that time many theories were proposed to explain this apparent complementarity, but the assumption of a solid Earth made these various proposals difficult to accept.

The discovery of radioactivity and its associated heating properties in 1895 prompted a re-examination of the apparent age of the Earth. This had previously been estimated by its cooling rate and assumption the Earth›s surface radiated like a black body. Those calculations had implied that, even if it started at red heat, the Earth would have dropped to its present temperature in a few tens of millions of years. Armed with the knowledge of a new heat source, scientists realized that the Earth would be much older, and that its core was still sufficiently hot to be liquid.

Alfred Wegener in Greenland in the winter of 1912-13.

By 1915, after having published a first article in 1912, Alfred Wegener was making serious arguments for the idea of continental drift in the first edition of *The Origin of Continents and Oceans*. In that book (re-issued in four successive editions up to the final one in 1936), he noted how the east coast of South America and the west coast of Africa looked as if they were once attached. Wegener was not the first to note this (Abraham Ortelius, Antonio Snider-Pellegrini, Eduard Suess, Roberto Mantovani and Frank Bursley Taylor preceded him just to mention a few), but he was the first to marshal significant fossil and paleo-topographical and climatological evidence to support this simple observation (and was supported in this by researchers such as Alex du Toit). Furthermore, when the rock strata of the margins of separate continents are very similar it suggests that these rocks were formed in the same way, implying that they were joined initially. For instance, parts of Scotland and Ireland contain rocks very similar to those found in Newfoundland and New Brunswick. Furthermore, the Caledonian Mountains of Europe and parts of the Appalachian Mountains of North America are very similar in structure and lithology.

However, his ideas were not taken seriously by many geologists, who pointed out that there was no apparent mechanism for continental drift. Specifically, they did not see how continental rock could plow through the much denser rock that makes up oceanic crust. Wegener could not explain the force that drove continental drift, and his vindication did not come until after his death in 1930.

Floating Continents, Paleomagnetism, and Seismicity Zones

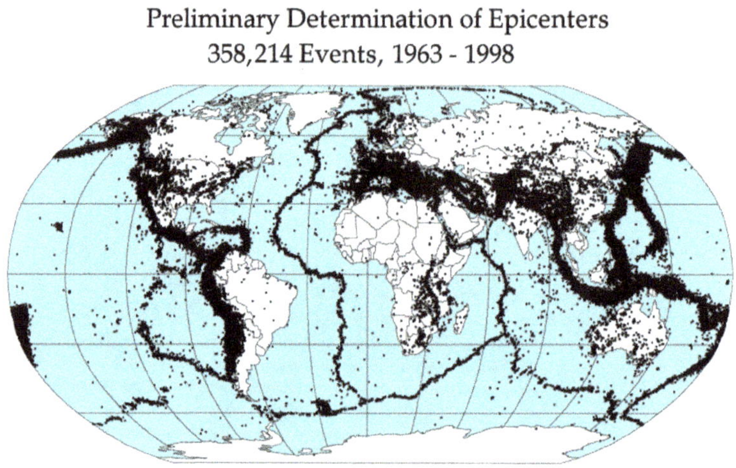

Global earthquake epicenters, 1963–1998.

As it was observed early that although granite existed on continents, seafloor seemed to be composed of denser basalt, the prevailing concept during the first half of the twentieth century was that there were two types of crust, named "sial" (continental type crust) and "sima" (oceanic type crust). Furthermore, it was supposed that a static shell of strata was present under the continents. It therefore looked apparent that a layer of basalt (sial) underlies the continental rocks.

However, based on abnormalities in plumb line deflection by the Andes in Peru, Pierre Bouguer had deduced that less-dense mountains must have a downward projection into the denser layer underneath. The concept that mountains had "roots" was confirmed by George B. Airy a hundred years later, during study of Himalayan gravitation, and seismic studies detected corresponding density variations. Therefore, by the mid-1950s, the question remained unresolved as to whether mountain roots were clenched in surrounding basalt or were floating on it like an iceberg.

During the 20th century, improvements in and greater use of seismic instruments such as seismographs enabled scientists to learn that earthquakes tend to be concentrated in specific areas, most notably along the oceanic trenches and spreading ridges. By the late 1920s, seismologists were beginning to identify several prominent earthquake zones parallel to the trenches that typically were inclined 40–60° from the horizontal and extended several hundred kilometers into the Earth. These zones later became known as Wadati-Benioff zones, or simply Benioff zones, in honor of the seismologists who first recognized them, Kiyoo Wadati of Japan and Hugo Benioff of the United States. The study of global seismicity greatly advanced in the 1960s with the establishment of the Worldwide Standardized Seismograph Network (WWSSN) to monitor the compliance of the 1963 treaty banning above-ground testing of nuclear weapons. The much improved data from the WWSSN instruments allowed seismologists to map precisely the zones of earthquake concentration worldwide. Meanwhile, debates developed around the phenomena of polar wander. Since the early debates of continental drift, scientists had discussed and used evidence that polar drift had occurred because continents seemed to have moved through different climatic zones during the past. Furthermore, paleomagnetic data had shown that the magnetic pole had also shifted during time. Reasoning in an opposite way, the continents might have shifted and rotated, while the pole remained relatively fixed. The first time the evidence of magnetic polar wander was used to support the movements of continents was in a paper by Keith Runcorn in 1956, and successive papers by him and his students Ted Irving (who was actually the first to be convinced of the fact that paleomagnetism supported continental drift) and Ken Creer.

This was immediately followed by a symposium in Tasmania in March 1956. In this symposium, the evidence was used in the theory of an expansion of the global crust. In this hypothesis the shifting of the continents can be simply explained by a large increase in size of the Earth since its formation. However, this was unsatisfactory because its supporters could offer no convincing mechanism to produce a significant expansion of the Earth. Certainly there is no evidence that the moon has expanded in the past 3 billion years; other work would soon show that the evidence was equally in support of continental drift on a globe with a stable radius.

During the thirties up to the late fifties, works by Vening-Meinesz, Holmes, Umbgrove, and numerous others outlined concepts that were close or nearly identical to modern plate tectonics theory. In particular, the English geologist Arthur Holmes proposed in 1920 that plate junctions might lie beneath the sea, and in 1928 that convection currents within the mantle might be the driving force. Often, these contributions are forgotten because:

- At the time, continental drift was not accepted.
- Some of these ideas were discussed in the context of abandoned fixistic ideas of a deforming globe without continental drift or an expanding Earth.
- They were published during an episode of extreme political and economic instability that hampered scientific communication.
- Many were published by European scientists and at first not mentioned or given little credit in the papers on sea floor spreading published by the American researchers in the 1960s.

Mid-oceanic Ridge Spreading and Convection

In 1947, a team of scientists led by Maurice Ewing utilizing the Woods Hole Oceanographic Institution›s research vessel *Atlantis* and an array of instruments, confirmed the existence of a rise in the central Atlantic Ocean, and found that the floor of the seabed beneath the layer of sediments consisted of basalt, not the granite which is the main constituent of continents. They also found that the oceanic crust was much thinner than continental crust. All these new findings raised important and intriguing questions.

The new data that had been collected on the ocean basins also showed particular characteristics regarding the bathymetry. One of the major outcomes of these datasets was that all along the globe, a system of mid-oceanic ridges was detected. An important conclusion was that along this system, new ocean floor was being created, which led to the concept of the "Great Global Rift". This was described in the crucial paper of Bruce Heezen (1960), which would trigger a real revolution in thinking. A profound consequence of seafloor spreading is that new crust was, and still is, being continually created along the oceanic ridges. Therefore, Heezen advocated the so-called "expanding Earth" hypothesis of S. Warren Carey. So, still the question remained: how can new crust be continuously added along the oceanic ridges without increasing the size of the Earth? In reality, this question had been solved already by numerous scientists during the forties and the fifties, like Arthur Holmes, Vening-Meinesz, Coates and many others: The crust in excess disappeared along what were called the oceanic trenches, where so-called "subduction" occurred. Therefore, when various scientists during the early sixties started to reason on the data at their disposal regarding the ocean floor, the pieces of the theory quickly fell into place.

The question particularly intrigued Harry Hammond Hess, a Princeton University geologist and a Naval Reserve Rear Admiral, and Robert S. Dietz, a scientist with the U.S. Coast and Geodetic Survey who first coined the term *seafloor spreading*. Dietz and Hess (the former published the same idea one year earlier in *Nature*, but priority belongs to Hess who had already distributed an unpublished manuscript of his 1962 article by 1960) were among the small handful who really understood the broad implications of sea floor spreading and how it would eventually agree with the, at that time, unconventional and unaccepted ideas of continental drift and the elegant and mobilistic models proposed by previous workers like Holmes.

In the same year, Robert R. Coats of the U.S. Geological Survey described the main features of island arc subduction in the Aleutian Islands. His paper, though little noted (and even ridiculed) at the time, has since been called "seminal" and "prescient". In reality, it actually shows that the work by the European scientists on island arcs and mountain belts performed and published during the 1930s up until the 1950s was applied and appreciated also in the United States.

If the Earth›s crust was expanding along the oceanic ridges, Hess and Dietz reasoned like Holmes and others before them, it must be shrinking elsewhere. Hess followed Heezen, suggesting that new oceanic crust continuously spreads away from the ridges in a conveyor belt–like motion. And, using the mobilistic concepts developed before, he correctly concluded that many millions of years later, the oceanic crust eventually descends along the continental margins where oceanic trenches – very deep, narrow canyons – are formed, e.g. along the rim of the Pacific Ocean basin. The important step Hess made was that convection currents would be the driving force in this process, arriving at the same conclusions as Holmes had decades before with the only difference that the thinning of the ocean crust was performed using Heezen›s mechanism of spreading along the ridges. Hess therefore concluded that the Atlantic Ocean was expanding while the Pacific Ocean was shrinking. As old oceanic crust is "consumed" in the trenches (like Holmes and others, he thought this was done by thickening of the continental lithosphere, not, as now understood, by underthrusting at a larger scale of the oceanic crust itself into the mantle), new magma rises and erupts along the spreading ridges to form new crust. In effect, the ocean basins are perpetually being "recycled," with the creation of new crust and the destruction of old oceanic lithosphere occurring simultaneously. Thus, the new mobilistic concepts neatly explained why the Earth does not get bigger with sea floor spreading, why there is so little sediment accumulation on the ocean floor, and why oceanic rocks are much younger than continental rocks.

Magnetic Striping

Beginning in the 1950s, scientists like Victor Vacquier, using magnetic instruments (magnetometers) adapted from airborne devices developed during World War II to detect submarines, began recognizing odd magnetic variations across the ocean floor. This finding, though unexpected, was not entirely surprising because it was known that basalt—the iron-rich, volcanic rock making up the ocean floor—contains a strongly magnetic mineral (magnetite) and can locally distort compass readings. This distortion was recognized by Icelandic mariners as early as the late 18th century. More important, because the presence of magnetite gives the basalt measurable magnetic properties, these newly discovered magnetic variations provided another means to study the deep ocean floor. When newly formed rock cools, such magnetic materials recorded the Earth›s magnetic field at the time.

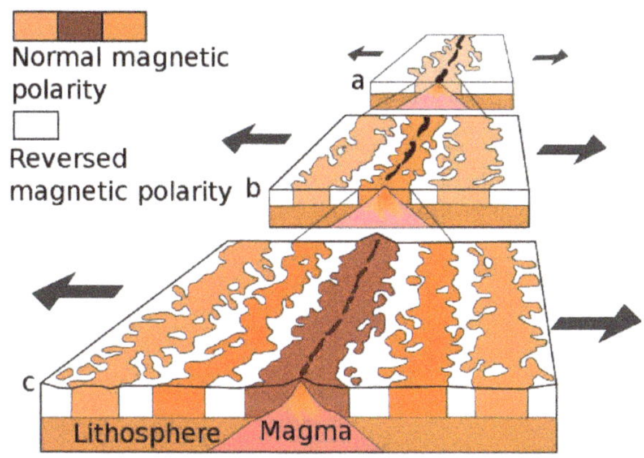

Seafloor magnetic striping.

A demonstration of magnetic striping. (The darker the color is, the closer it is to normal polarity).

As more and more of the seafloor was mapped during the 1950s, the magnetic variations turned out not to be random or isolated occurrences, but instead revealed recognizable patterns. When these magnetic patterns were mapped over a wide region, the ocean floor showed a zebra-like pattern: one stripe with normal polarity and the adjoining stripe with reversed polarity. The overall pattern, defined by these alternating bands of normally and reversely polarized rock, became known as magnetic striping, and was published by Ron G. Mason and co-workers in 1961, who did not find, though, an explanation for these data in terms of sea floor spreading, like Vine, Matthews and Morley a few years later.

The discovery of magnetic striping called for an explanation. In the early 1960s scientists such as Heezen, Hess and Dietz had begun to theorise that mid-ocean ridges mark structurally weak zones where the ocean floor was being ripped in two lengthwise along the ridge crest (see the previous paragraph). New magma from deep within the Earth rises easily through these weak zones and eventually erupts along the crest of the ridges to create new oceanic crust. This process, at first denominated the "conveyer belt hypothesis" and later called seafloor spreading, operating over many millions of years continues to form new ocean floor all across the 50,000 km-long system of mid-ocean ridges.

Only four years after the maps with the "zebra pattern" of magnetic stripes were published, the link between sea floor spreading and these patterns was correctly placed, independently by Lawrence Morley, and by Fred Vine and Drummond Matthews, in 1963, now called the Vine-Matthews-Morley hypothesis. This hypothesis linked these patterns to geomagnetic reversals and was supported by several lines of evidence:

- the stripes are symmetrical around the crests of the mid-ocean ridges; at or near the crest of the ridge, the rocks are very young, and they become progressively older away from the ridge crest;
- the youngest rocks at the ridge crest always have present-day (normal) polarity;
- stripes of rock parallel to the ridge crest alternate in magnetic polarity (normal-reversed-normal, etc.), suggesting that they were formed during different epochs documenting the (already known from independent studies) normal and reversal episodes of the Earth›s magnetic field.

By explaining both the zebra-like magnetic striping and the construction of the mid-ocean ridge system, the seafloor spreading hypothesis (SFS) quickly gained converts and represented another major advance in the development of the plate-tectonics theory. Furthermore, the oceanic crust now came to be appreciated as a natural "tape recording" of the history of the geomagnetic field reversals (GMFR) of the Earth›s magnetic field. Today, extensive studies are dedicated to the calibration of the normal-reversal patterns in the oceanic crust on one hand and known timescales derived from the dating of basalt layers in sedimentary sequences (magnetostratigraphy) on the other, to arrive at estimates of past spreading rates and plate reconstructions.

Definition and Refining of the Theory

After all these considerations, Plate Tectonics (or, as it was initially called "New Global Tectonics") became quickly accepted in the scientific world, and numerous papers followed that defined the concepts:

- In 1965, Tuzo Wilson who had been a promotor of the sea floor spreading hypothesis and continental drift from the very beginning added the concept of transform faults to the model, completing the classes of fault types necessary to make the mobility of the plates on the globe work out.
- A symposium on continental drift was held at the Royal Society of London in 1965 which must be regarded as the official start of the acceptance of plate tectonics by the scientific community, and which abstracts are issued as Blacket, Bullard & Runcorn (1965). In this symposium, Edward Bullard and co-workers showed with a computer calculation how the continents along both sides of the Atlantic would best fit to close the ocean, which became known as the famous "Bullard's Fit".
- In 1966 Wilson published the paper that referred to previous plate tectonic reconstructions,

introducing the concept of what is now known as the "Wilson Cycle".

- In 1967, at the American Geophysical Union›s meeting, W. Jason Morgan proposed that the Earth's surface consists of 12 rigid plates that move relative to each other.
- Two months later, Xavier Le Pichon published a complete model based on 6 major plates with their relative motions, which marked the final acceptance by the scientific community of plate tectonics.
- In the same year, McKenzie and Parker independently presented a model similar to Morgan's using translations and rotations on a sphere to define the plate motions.

IMPLICATIONS FOR BIOGEOGRAPHY

Continental drift theory helps biogeographers to explain the disjunct biogeographic distribution of present-day life found on different continents but having similar ancestors. In particular, it explains the Gondwanan distribution of ratites and the Antarctic flora.

PLATE RECONSTRUCTION

Reconstruction is used to establish past (and future) plate configurations, helping determine the shape and make-up of ancient supercontinents and providing a basis for paleogeography.

Defining Plate Boundaries

Current plate boundaries are defined by their seismicity. Past plate boundaries within existing plates are identified from a variety of evidence, such as the presence of ophiolites that are indicative of vanished oceans.

Past Plate Motions

Tectonic motion first began around three billion years ago.

Various types of quantitative and semi-quantitative information are available to constrain past plate motions. The geometric fit between continents, such as between west Africa and South America is still an important part of plate reconstruction. Magnetic stripe patterns provide a reliable guide to relative plate motions going back into the Jurassic period. The tracks of hotspots give absolute reconstructions, but these are only available back to the Cretaceous. Older reconstructions rely mainly on paleomagnetic pole data, although these only constrain the latitude and rotation, but not the longitude. Combining poles of different ages in a particular plate to produce apparent polar wander paths provides a method for comparing the motions of different plates through time. Additional evidence comes from the distribution of certain sedimentary rock types, faunal provinces shown by particular fossil groups, and the position of orogenic belts.

Formation and Break-up of Continents

The movement of plates has caused the formation and break-up of continents over time, including occasional formation of a supercontinent that contains most or all of the continents. The supercontinent Columbia or Nuna formed during a period of 2,000 to 1,800 million years ago and broke up about 1,500 to 1,300 million years ago. The supercontinent Rodinia is thought to have formed about 1 billion years ago and to have embodied most or all of Earth's continents, and broken up into eight continents around 600 million years ago. The eight continents later re-assembled into another supercontinent called Pangaea; Pangaea broke up into Laurasia (which became North America and Eurasia) and Gondwana (which became the remaining continents).

The Himalayas, the world›s tallest mountain range, are assumed to have been formed by the collision of two major plates. Before uplift, they were covered by the Tethys Ocean.

CURRENT PLATES

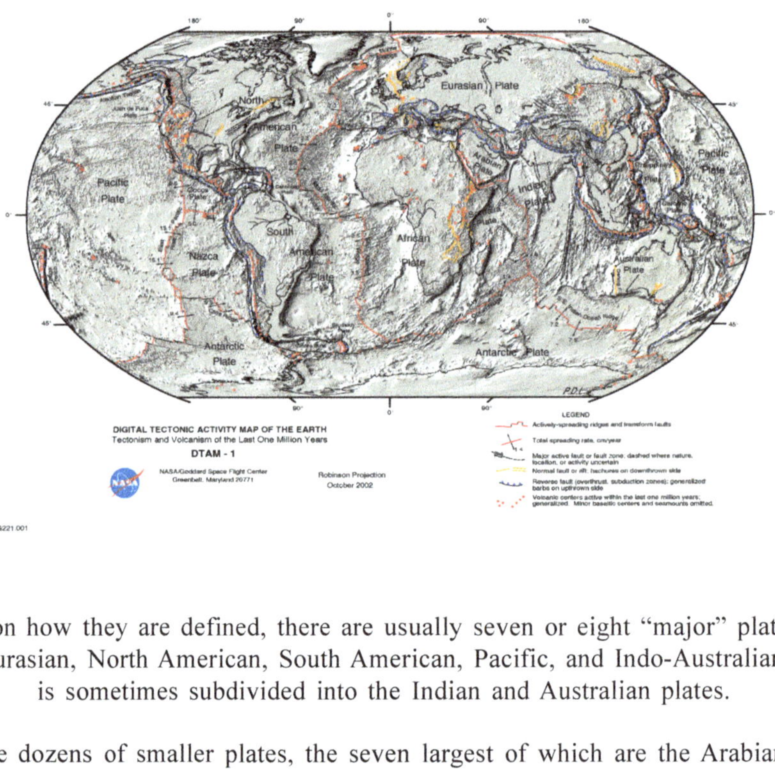

Depending on how they are defined, there are usually seven or eight "major" plates: African, Antarctic, Eurasian, North American, South American, Pacific, and Indo-Australian. The latter is sometimes subdivided into the Indian and Australian plates.

There are dozens of smaller plates, the seven largest of which are the Arabian, Caribbean, Juan de Fuca, Cocos, Nazca, Philippine Sea and Scotia.

The current motion of the tectonic plates is today determined by remote sensing satellite data sets, calibrated with ground station measurements.

Other celestial bodies (planets, moons)

The appearance of plate tectonics on terrestrial planets is related to planetary mass, with more massive planets than Earth expected to exhibit plate tectonics. Earth may be a borderline case, owing its tectonic activity to abundant water (silica and water form a deep eutectic.)

Venus

Venus shows no evidence of active plate tectonics. There is debatable evidence of active tectonics in the planet's distant past; however, events taking place since then (such as the plausible and generally accepted hypothesis that the Venusian lithosphere has thickened greatly over the course of several hundred million years) has made constraining the course of its geologic record difficult. However, the numerous well-preserved impact craters have been utilized as a dating method to approximately date the Venusian surface (since there are thus far no known samples of Venusian rock to be dated by more reliable methods). Dates derived are dominantly in the range 500 to 750 million years ago, although ages of up to 1,200 million years ago have been calculated. This research has led to the fairly well accepted hypothesis that Venus has undergone an essentially complete volcanic resurfacing at least once in its distant past, with the last event taking place approximately within the range of estimated surface ages. While the mechanism of such an impressive thermal event remains a debated issue in Venusian geosciences, some scientists are advocates of processes involving plate motion to some extent.

One explanation for Venus›s lack of plate tectonics is that on Venus temperatures are too high for significant water to be present. The Earth›s crust is soaked with water, and water plays an important role in the development of shear zones. Plate tectonics requires weak surfaces in the crust along which crustal slices can move, and it may well be that such weakening never took place on Venus because of the absence of water. However, some researchers remain convinced that plate tectonics is or was once active on this planet.

Mars

Mars is considerably smaller than Earth and Venus, and there is evidence for ice on its surface and in its crust.

In the 1990s, it was proposed that Martian Crustal Dichotomy was created by plate tectonic processes. Scientists today disagree, and think that it was created either by upwelling within the Martian mantle that thickened the crust of the Southern Highlands and formed Tharsis or by a giant impact that excavated the Northern Lowlands.

Valles Marineris may be a tectonic boundary.

Observations made of the magnetic field of Mars by the *Mars Global Surveyor* spacecraft in 1999 showed patterns of magnetic striping discovered on this planet. Some scientists interpreted these as requiring plate tectonic processes, such as seafloor spreading. However, their data fail a "magnetic reversal test", which is used to see if they were formed by flipping polarities of a global magnetic field.

Icy Satellites

Some of the satellites of Jupiter have features that may be related to plate-tectonic style deformation, although the materials and specific mechanisms may be different from plate-tectonic activity on Earth. On 8 September 2014, NASA reported finding evidence of plate tectonics on Europa, a satellite of Jupiter—the first sign of such geological activity on another world other than Earth.

Titan, the largest moon of Saturn, was reported to show tectonic activity in images taken by the *Huygens* probe, which landed on Titan on January 14, 2005.

Exoplanets

On Earth-sized planets, plate tectonics is more likely if there are oceans of water; however, in 2007, two independent teams of researchers came to opposing conclusions about the likelihood of plate tectonics on larger super-earths with one team saying that plate tectonics would be episodic or stagnant and the other team saying that plate tectonics is very likely on super-earths even if the planet is dry.

REFERENCES

1. Little, Fowler & Coulson 1990.
2. Read & Watson 1975.
3. Scalera & Lavecchia 2006.
4. Zhen Shao 1997, Hancock, Skinner & Dineley 2000.
5. Turcotte & Schubert 2002, p. 5.
6. Turcotte & Schubert 2002.
7. Turcotte & Schubert 2002, p. 3.
8. Foulger 2010.
9. Schmidt & Harbert 1998.
10. Meissner 2002, p. 100.
11. Plate Tectonics: Plate Boundaries". platetectonics.com. Retrieved 12 June 2010.
12. Understanding plate motions". USGS. Retrieved 12 June 2010.
13. Grove, Timothy L.; Till, Christy B.; Krawczynski, Michael J. (8 March 2012). "The Role of H2O in Subduction Zone Magmatism". Annual Review of Earth and Planetary Sciences (40): 413–439. Bibcode:2012AREPS..40..413G. doi:10.1146/annurev-earth-042711-105310. Retrieved 14 January 2016.
14. Mendia-Landa, Pedro. "Myths and Legends on Natural Disasters: Making Sense of Our World". Retrieved 2008-02-05.
15. van Dijk 1992, van Dijk & Okkes 1991.
16. Holmes, Arthur (1931). "Radioactivity and Earth Movements" (PDF). Transactions of the Geological Society of Glasgow. Geological Society of Glasgow: 559–606.
17. Tanimoto & Lay 2000.
18. Meyerhoff et al. 1996.
19. Mallard, Claire; Coltice, Nicolas; Seton, Maria; Müller, R. Dietmar; Tackley, Paul J. (2016). "Subduction controls the distribution and fragmentation of Earth›s tectonic plates". Nature. 535 *(7610): 140–143. doi:10.1038/nature17992. ISSN 0028-0836.*
20. Spence 1987, White & McKenzie 1989.
21. Conrad & Lithgow-Bertelloni 2002.

22. Spence 1987, White & Mckenzie 1989, Segev 2002.
23. Alfred Wegener (1880-1930)". University of California Museum of Paleontology.
24. Neith, Katie (April 15, 2011). "Caltech Researchers Use GPS Data to Model Effects of Tidal Loads on Earth›s Surface". Caltech. Retrieved August 15, 2012.
25. Ricard, Y. (2009). "2. Physics of Mantle Convection". In David Bercovici; Gerald Schubert. Treatise on Geophysics: Mantle Dynamics. 7. Elsevier Science. p. 36.
26. Glatzmaier, Gary A. (2013). Introduction to Modeling Convection in Planets and Stars: Magnetic Field, Density Stratification, Rotation. Princeton University Press. p. 149.
27. van Dijk 1992, van Dijk & Okkes 1990.
28. Moore 1973.
29. Bostrom 1971.
30. Scoppola et al. 2006.
31. Torsvik et al. 2010.
32. Wegener 1929.
33. Hughes 2001a.
34. Wegener 1966, Hughes 2001b.
35. Runcorn 1956.
36. Carey 1956.
37. see for example the milestone paper of Lyman & Fleming 1940.
38. Korgen 1995, Spiess & Kuperman 2003.
39. Kious & Tilling 1996.
40. Frankel 1987.
41. Joly 1909.
42. Thomson 1863.
43. Wegener 1912.
44. Stein & Wysession 2009, p. 26
45. Carey 1956; see also Quilty 2003.
46. Holmes 1928; see also Holmes 1978, Frankel 1978.
47. Lippsett 2001, Lippsett 2006.
48. Heezen 1960.
49. Dietz 1961.
50. Hess 1962.
51. Mason & Raff 1961, Raff & Mason 1961.
52. Vine & Matthews 1963.
53. See summary in Heirzler, Le Pichon & Baron 1966
54. Wilson 1963.
55. Wilson 1965.

56. Wilson 1966.
57. Morgan 1968.
58. Le Pichon 1967.
59. McKenzie & Parker 1967.
60. Moss & Wilson 1998.
61. Condie 1997.
62. Lliboutry 2000.
63. Kranendonk, V.; Martin, J. (2011). "Onset of Plate Tectonics". Science. 333 *(6041): 413–414. Bibcode:2011Sci...333..413V. doi:10.1126/science.1208766. PMID 21778389.*
64. Torsvik, Trond Helge. "Reconstruction Methods". Retrieved 18 June 2010.
65. Torsvik 2008.
66. Butler 1992.
67. Scotese, C.R. (2002-04-20). "Climate History". Paleomap Project. Retrieved 18 June 2010.
68. Zhao 2002, 2004
69. Valencia, O'Connell & Sasselov 2007.
70. Kasting 1988.
71. Bortman, Henry (2004-08-26). "Was Venus alive? "The Signs are Probably There"". Astrobiology Magazine. Retrieved 2008-01-08.
72. Sleep 1994.
73. Zhong & Zuber 2001.
74. Andrews-Hanna, Zuber & Banerdt 2008.
75. Wolpert, Stuart (August 9, 2012). "UCLA scientist discovers plate tectonics on Mars". Yin, An. UCLA. Retrieved August 13, 2012.
76. Connerney et al. 1999, Connerney et al. 2005
77. Harrison 2000.
78. Dyches, Preston; Brown, Dwayne; Buckley, Michael (8 September 2014). "Scientists Find Evidence of ‹Diving› Tectonic Plates on Europa". NASA. Retrieved 8 September 2014.
79. Soderblom et al. 2007.
80. Valencia, Diana; O›Connell, Richard J. "Convection scaling and subduction on Earth and super-Earths". Earth and Planetary Science Letters. 286 *(3–4): 492–502. Bibcode:2009E&PSL.286..492V. doi:10.1016/j.epsl.2009.07.015.*
81. van Heck, H. J.; Tackley, P.J. "Plate tectonics on super-Earths: Equally or more likely than on Earth". Earth and Planetary Science Letters. 310 *(3–4): 252–261. Bibcode:2011E&PSL.310..252V. doi:10.1016/j.epsl.2011.07.029.*
82. C. O›Neill, A. Lenardic Geological consequences of super-sized Earths GEOPHYSICAL RESEARCH LETTERS 34: L19204 doi:10.1029/2007GL030598

CHAPTER 3

Mid-ocean Ridge History and Formation Processes

Chapter Contents

The mid-ocean ridge is the world's longest mountain range, although it is submerged. It is a continuous ocean ridge that stretches through all the world's oceans, including the Pacific, Atlantic, Indian, Arctic, and Southern Ocean. It exists along the boundary of tectonic plates. In fact, the only reason that the mid-ocean ridge exists is due to weaknesses in the Earth's crust at plate interfaces which permit volcanic activity. The volcanic activity generates magma, which cools to form this continuous submarine mountain range.

INTRODUCTION

A **mid-ocean ridge** is an underwater mountain system formed by plate tectonics. It consists of various mountains linked in chains, typically having a valley known as a rift running along its spine. This type of oceanic mountain ridge is characteristic of what is known as an oceanic spreading center, which is responsible for seafloor spreading. The production of new seafloor results from mantle upwelling in response to plate spreading; this isentropic upwelling solid mantle material eventually exceeds the solidus and melts. The buoyant melt rises as magma at a linear weakness in the oceanic crust, and emerges as lava, creating new crust upon cooling. A mid-ocean ridge demarcates the boundary between two tectonic plates, and consequently is termed a divergent plate boundary.

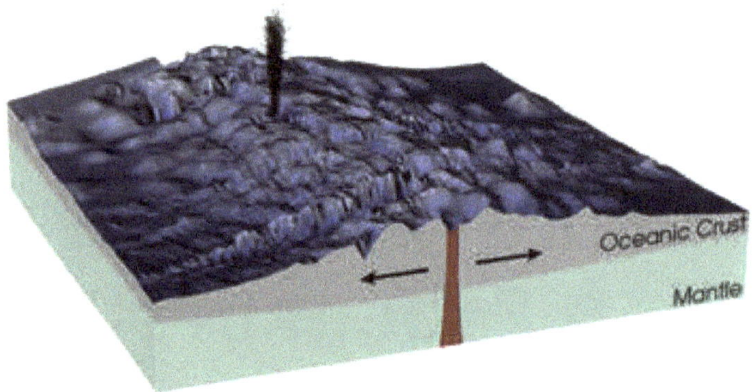

Mid-oceanic ridge, including a black smoker

Description

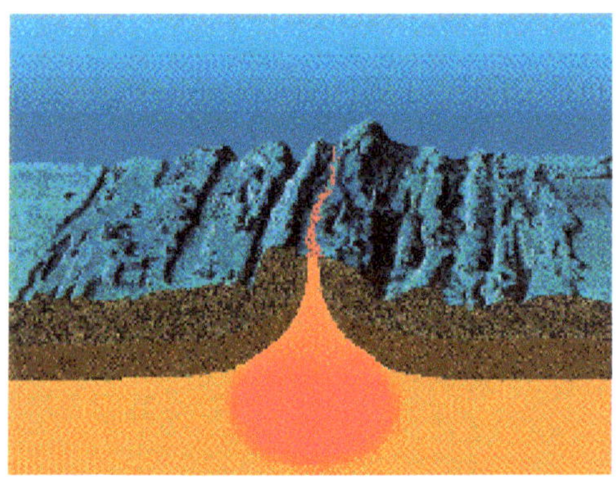

A close-up showing a mid-ocean ridge topography with magma rising from a chamber below, forming new ocean plate which spreads away from ridge.

Mid-ocean ridges are geologically active, with new magma constantly emerging onto the ocean floor and into the crust at and near rifts along the ridge axes. The crystallized magma forms new crust of basalt (known as MORB for mid-ocean ridge basalt) and gabbro. They are formed by two oceanic plates moving away from each other.

The rocks making up the crust below the seafloor are youngest at the axis of the ridge and age with increasing distance from that axis. New magma of basalt composition emerges at and near the axis because of decompression melting in the underlying Earth's mantle.

The oceanic crust is made up of rocks much younger than the Earth itself. Most oceanic crust in the ocean basins is less than 200 million years old. The crust is in a constant state of "renewal" at the ocean ridges. Moving away from the mid-ocean ridge, ocean depth progressively increases; the greatest depths are in ocean trenches. As the oceanic crust moves away from the ridge axis, the peridotite in the underlying mantle cools and becomes more rigid. The crust and the relatively rigid peridotite below it make up the oceanic lithosphere.

Slow spreading ridges like the Mid-Atlantic Ridge (MAR) generally have large, wide rift valleys, sometimes as wide as 10–20 km (6.2–12.4 mi), and very rugged terrain at the ridge crest that can have relief of up to a 1,000 m (3,300 ft). By contrast, fast spreading ridges like the East Pacific Rise (EPR) are narrow, sharp incisions surrounded by generally flat topography that slopes away from the ridge over many hundreds of miles.

The overall shape of ridges results from Pratt isostacy: close to the ridge axis there is hot, low-density mantle supporting the oceanic crust. As the oceanic plates cool, away from the ridge axes, the oceanic mantle lithosphere (the colder, denser part of the mantle that, together with the crust, comprises the oceanic plates) thickens and the density increases. Thus older seafloor is underlain by denser material and 'sits' lower. The width of the ridge is hence a function of spreading rate - slow ridges like the MAR have spread much less far than faster ridges like the EPR for the same amount of cooling and consequent bathymetric drop-off.

FORMATION PROCESSES

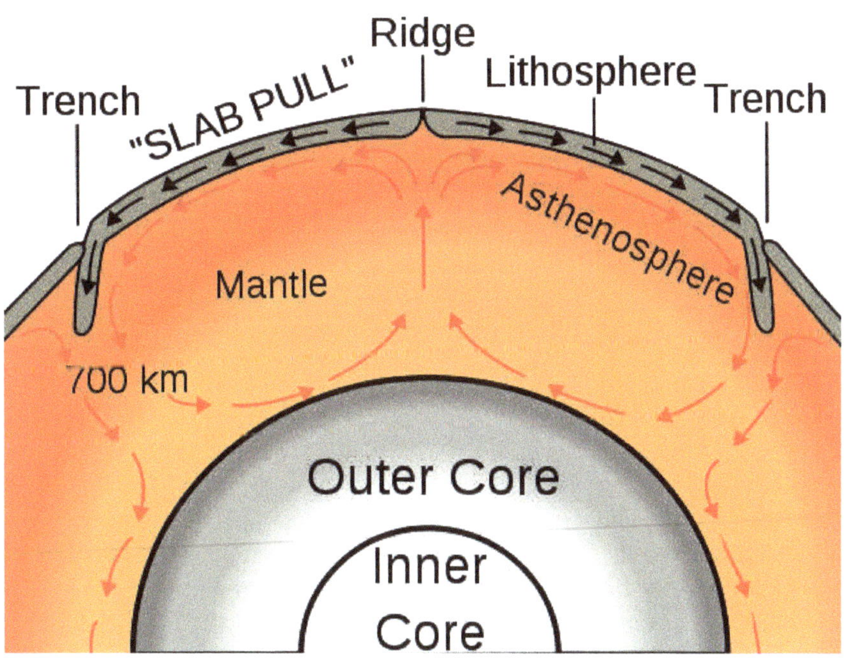

Oceanic crust is formed at an oceanic ridge, while the lithosphere is subducted back into the asthenosphere at trenches.

There are two processes, ridge-push and slab pull, thought to be responsible for the spreading seen at mid-ocean ridges, and there is some uncertainty as to which is dominant. Ridge-push occurs when the growing bulk of the ridge pushes the rest of the tectonic plate away from the ridge, often towards a subduction zone. At the subduction zone, slab-pull comes into effect. This is simply the weight of the tectonic plate being subducted (pulled) below the overlying plate dragging the rest of the plate along behind it.

The other process proposed to contribute to the formation of new oceanic crust at mid-ocean ridges is the mantle conveyor (see image). However, there have been some studies which have shown that the upper mantle (asthenosphere) is too plastic (flexible) to generate enough friction to pull the tectonic plate along. Moreover, unlike in the image above, mantle upwelling that causes magma to form beneath the ocean ridges appears to involve only its upper 400 km (250 mi), as deduced from seismic tomography and from studies of the seismic discontinuity at about 400 km (250 mi). The relatively shallow depths from which the upwelling mantle rises below ridges are more consistent with the "slab-pull" process. On the other hand, some of the world's largest tectonic plates such as the North American Plate are in motion, yet are nowhere being subducted.

The rate at which the mid-ocean ridge creates new material is known as the spreading rate, and is generally measured in mm/yr. The common subdivisions of spreading rate are fast, medium, and slow with values generally being >100 mm/yr, 100–55 mm/yr, and 55–20 mm/yr, respectively. The spreading rate of the North Atlantic Ocean is ~ 25 mm/yr, while in the Pacific region, it is 80–120 mm/yr. Ridges that spread at rates <20 mm/yr are referred to as ultraslow spreading ridges (e.g., the Gakkel Ridge in the Arctic Ocean and the Southwest Indian Ridge) and they provide a much different perspective on crustal formation than their faster spreading brethren.

The mid-ocean ridge systems form new oceanic crust. As crystallized basalt extruded at a ridge axis cools below Curie points of appropriate iron-titanium oxides, magnetic field directions parallel to the Earth's magnetic field are recorded in those oxides. The orientations of the field in the oceanic crust record preserve a record of directions of the Earth's magnetic field with time. Because the field has reversed directions at irregular intervals throughout its history, the pattern of geomagnetic reversals in the ocean crust can be used as an indicator of age. Likewise, the pattern of reversals together with age measurements of the crust is used to help establish the history of the Earth's magnetic field.

Global System

The mid-ocean ridges of the world are connected and form *the* Ocean Ridge, a single global mid-oceanic ridge system that is part of every ocean, making it the longest mountain range in the world. The continuous mountain range is 65,000 km (40,400 mi) long (several times longer than the Andes, the longest continental mountain range), and the total length of the oceanic ridge system is 80,000 km (49,700 mi) long.

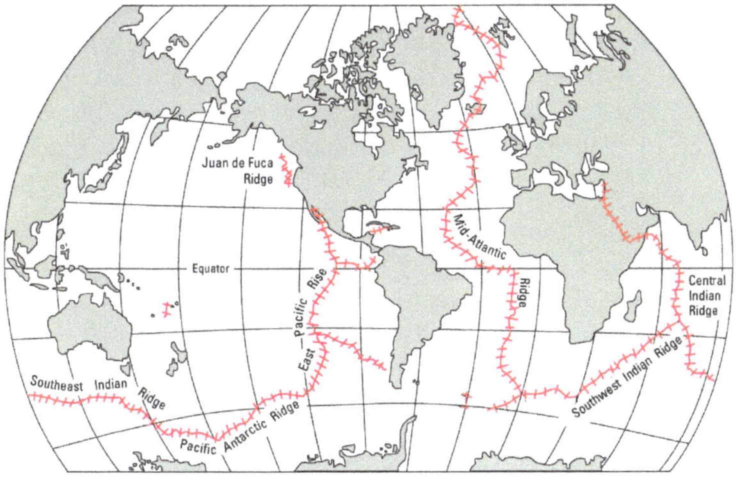

World Distribution of Mid-Oceanic Ridges; USGS.

HISTORY

Discovery

Mid-ocean ridges are generally submerged deep in the ocean. It was not until the 1950s, when the ocean floor was surveyed in detail, that their full extent became known.

The *Vema*, a ship of the Lamont-Doherty Earth Observatory of Columbia University, traversed the Atlantic Ocean, recording data about the ocean floor from the ocean surface. A team led by Marie Tharp and Bruce Heezen analyzed the data and concluded that there was an enormous mountain chain running up the middle. Scientists named it the "Mid-Atlantic Ridge.

At first, the ridge was thought to be a phenomenon specific to the Atlantic Ocean. However, as surveys of the ocean floor continued around the world, it was discovered that every ocean contains parts of the mid-ocean ridge system. Although the ridge system runs down the middle of the Atlantic Ocean, the ridge is located away from the center of other oceans.

Impact

Alfred Wegener proposed the theory of continental drift in 1912. He stated: "the Mid-Atlantic Ridge ... zone in which the floor of the Atlantic, as it keeps spreading, is continuously tearing open and making space for fresh, relatively fluid and hot sima [rising] from depth." However, Wegener did not pursue this observation in his later works and his theory was dismissed by geologists because there was no mechanism to explain how continents could plow through ocean crust, and the theory became largely forgotten.

Following the discovery of the worldwide extent of the mid-ocean ridge in the 1950s, geologists faced a new task: explaining how such an enormous geological structure could have formed. In the 1960s, geologists discovered and began to propose mechanisms for seafloor spreading. Plate tectonics was a suitable explanation for seafloor spreading, and the acceptance of plate tectonics by the majority of geologists resulted in a major paradigm shift in geological thinking.

It is estimated that 20 volcanic eruptions occur each year along earth's mid-ocean ridges and that every year 2.5 km^2(0.97 sq mi) of new seafloor is formed by this process. With a crustal thickness of 1 to 2 km (0.62 to 1.24 mi), this amounts to about 4 km^3 (0.96 cu mi) of new ocean crust formed every year.

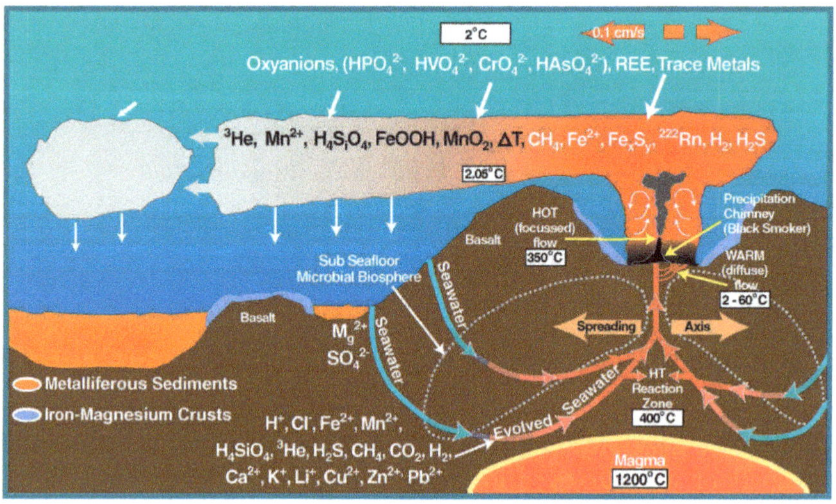

Oceanic ridge and deep sea vent chemistry

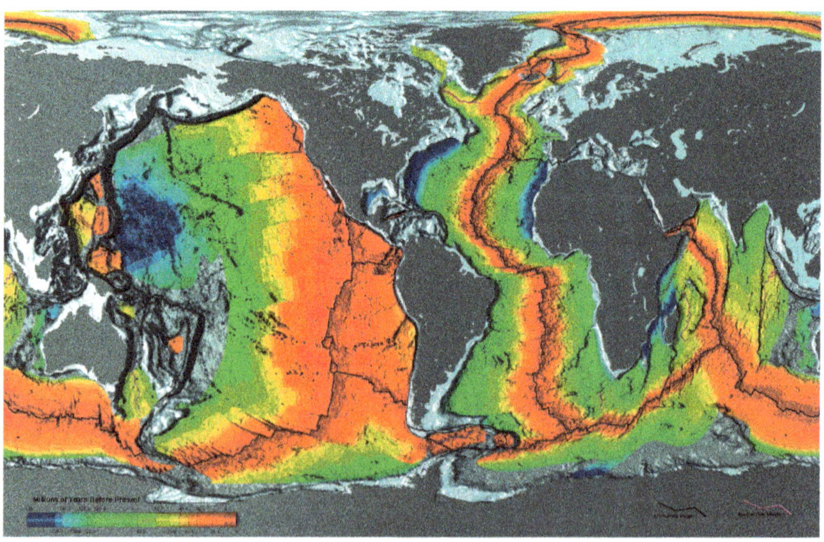

Age of oceanic crust. The red is most recent, and blue is the oldest.

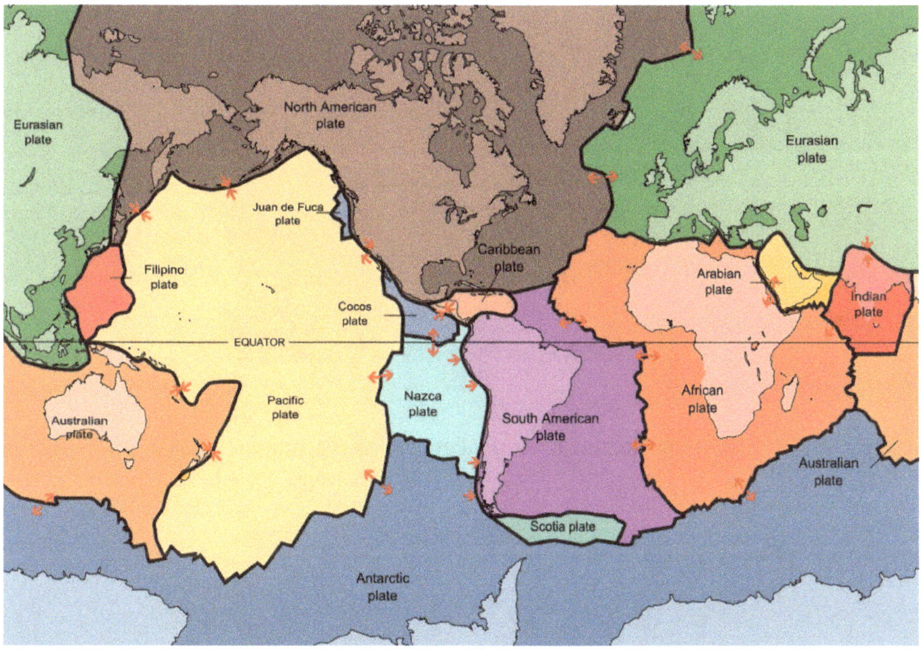

Plates in the crust of the earth, according to the plate tectonics theory.

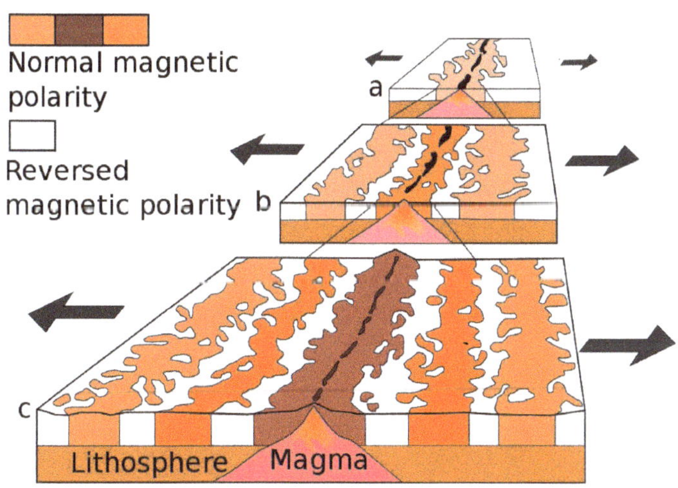

Seafloor magnetic striping.

A demonstration of magnetic striping..

HOTSPOT (GEOLOGY)

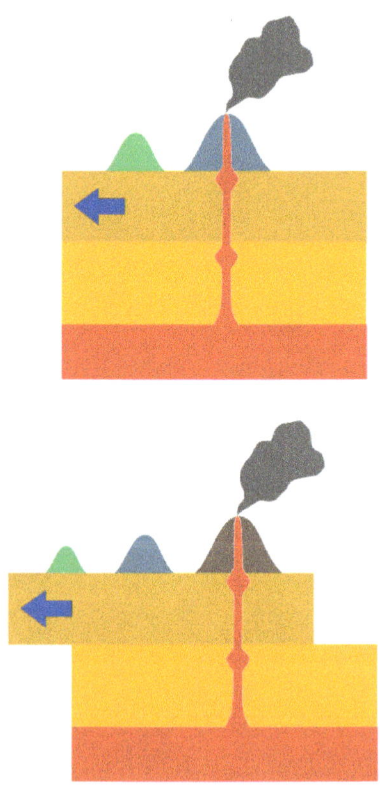

Diagram showing a cross section though the Earth's lithosphere (in yellow) with magma rising from the mantle (in red).

In geology, the places known as **hotspots** or **hot spots** are volcanic regions thought to be fed by underlying mantle that is anomalously hot compared with the surrounding mantle. They may be on, near to, or far from tectonic plate boundaries. Currently, there are two hypotheses that attempt to explain their origins. One suggests that hotspots are due to mantle plumes that rise as thermal diapirs from the core–mantle boundary. The other hypothesis is that lithospheric extension permits the passive rising of melt from shallow depths. This hypothesis considers the term "hotspot" to be a misnomer, asserting that the mantle source beneath them is, in fact, not anomalously hot at all. Well known examples include Hawaii and Yellowstone.

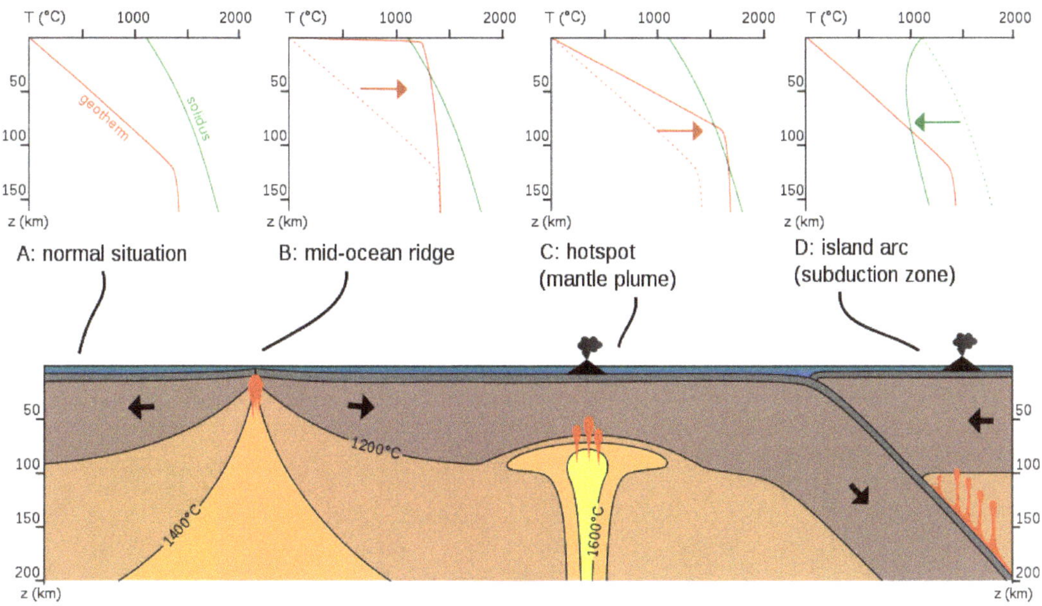

Schematic diagram showing the physical processes inside the Earth that lead to the generation of magma. Partial melting begins above the fusion point.

The origins of the concept of hotspots lie in the work of J. Tuzo Wilson, who postulated in 1963 that the Hawaiian Islands result from the slow movement of a tectonic plate across a hot region beneath the surface. It was later postulated that hotspots are fed by narrow streams of hot mantle rising from the Earth's core–mantle boundary in a structure called a mantle plume. Whether or not such mantle plumes exist is currently the subject of a major controversy in Earth science. Estimates for the number of hotspots postulated to be fed by mantle plumes has ranged from about 20 to several thousands, over the years, with most geologists considering a few tens to exist. Hawaii, Réunion, Yellowstone, Galápagos, and Iceland are some of the currently most active volcanic regions to which the hypothesis is applied.

Most hotspot volcanoes are basaltic (e.g., Hawaii, Tahiti). As a result, they are less explosive than subduction zone volcanoes, in which water is trapped under the overriding plate. Where hotspots occur in continental regions, basalticmagma rises through the continental crust, which melts to form rhyolites. These rhyolites can form violent eruptions. For example, the Yellowstone Caldera was formed by some of the most powerful volcanic explosions in geologic history. However, when the rhyolite is completely erupted, it may be followed by eruptions of basaltic magma rising through the same lithospheric fissures (cracks in the lithosphere). An example of this activity is the Ilgachuz Range in British Columbia, which was created by an early complex series of trachyte and rhyolite eruptions, and late extrusion of a sequence of basaltic lava flows.

The hotspot hypothesis is now closely linked to the mantle plume hypothesis.

Comparison with Island arc Volcanoes

Hotspot volcanoes are considered to have a fundamentally different origin from island arc volcanoes. The latter form over subduction zones, at converging plate boundaries. When one oceanic plate meets another, the denser plate is forced downward into a deep ocean trench. This plate, as it is subducted, releases water into the base of the over-riding plate, and this water mixes with the rock, thus changing its composition causing some rock to melt and rise. It is this that fuels a chain of volcanoes, such as the Aleutian Islands, near Alaska.

Hotspot Volcanic Chains

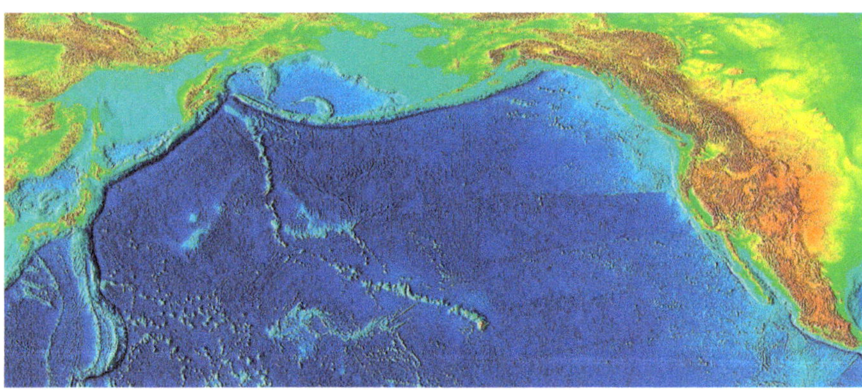

Over millions of years, the Pacific Plate has moved over the Hawaii hotspot, creating a trail of underwater mountains that stretch across the Pacific.

Kilauea is the most active shield volcano in the world. The volcano has erupted nonstop since 1983 and it is part of the Hawaiian–Emperor seamount chain.

Mauna Loa is a large shield volcano. Its last eruption was in 1984 and it is part of the Hawaiian–Emperor seamount chain.

Bowie Seamount is a dormant submarine volcano and it is part of the Kodiak-Bowie Seamount chain.

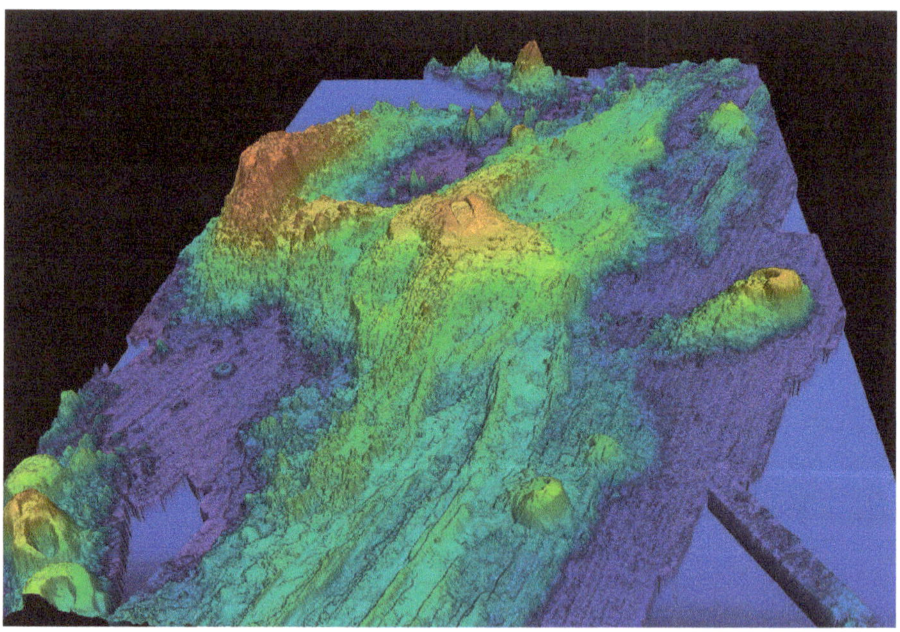

Axial Seamount is the youngest seamount of the Cobb–Eickelberg Seamount chain. Its last eruption was on 6 April 2011.

Mauna Kea is the tallest volcano in the Hawaiian–Emperor seamount chain. It is currently dormant and it has cinder cones growing on the volcano.

Hualalai is a massive shield volcano in the Hawaiian–Emperor seamount chain. Its last eruption was in 1801.

The joint mantle plume/hotspot hypothesis envisages the feeder structures to be fixed relative to one another, with the continents and seafloor drifting overhead. The hypothesis thus predicts that time-progressive chains of volcanoes are developed on the surface. Examples are Yellowstone, which lies at the end of a chain of extinct calderas, which become progressively older to the west. Another example is the Hawaiian archipelago, where islands become progressively older and more deeply eroded to the northwest.

Geologists have tried to use hotspot volcanic chains to track the movement of the Earth's tectonic plates. This effort has been vexed by the lack of very long chains, by the fact that many are not time-progressive (e.g. the Galápagos) and by the fact that hotspots do not appear to be fixed relative to one another (e.g. Hawaii and Iceland.)

Postulated Hotspot Volcano Chains

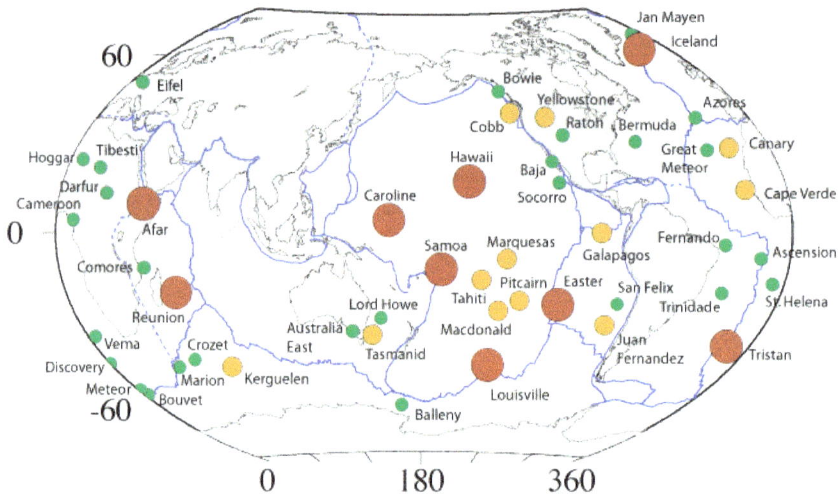

An example of mantle plume locations suggested by one recent group. Figure from Foulger (2010).

- Hawaiian–Emperor seamount chain (Hawaii hotspot)
- Louisville seamount chain (Louisville hotspot)
- Walvis Ridge (Gough and Tristan hotspot)
- Kodiak–Bowie Seamount chain (Bowie hotspot)
- Cobb–Eickelberg Seamount chain (Cobb hotspot)
- New England Seamount chain (New England hotspot)
- Anahim Volcanic Belt (Anahim hotspot)
- Mackenzie dike swarm (Mackenzie hotspot)
- Great Meteor hotspot track (New England hotspot)
- St. Helena Seamount Chain – Cameroon Volcanic Line (Saint Helena hotspot)
- Southern Mascarene Plateau–Chagos-Maldives-Laccadive Ridge (Réunion hotspot)
- Ninety East Ridge (Kerguelen hotspot)
- Tuamotu–Line Island chain (Easter hotspot)
- Austral–Gilbert–Marshall chain (Macdonald hotspot)
- Juan Fernández Ridge (Juan Fernández hotspot)

List of Volcanic Regions Postulated to be Hotspots

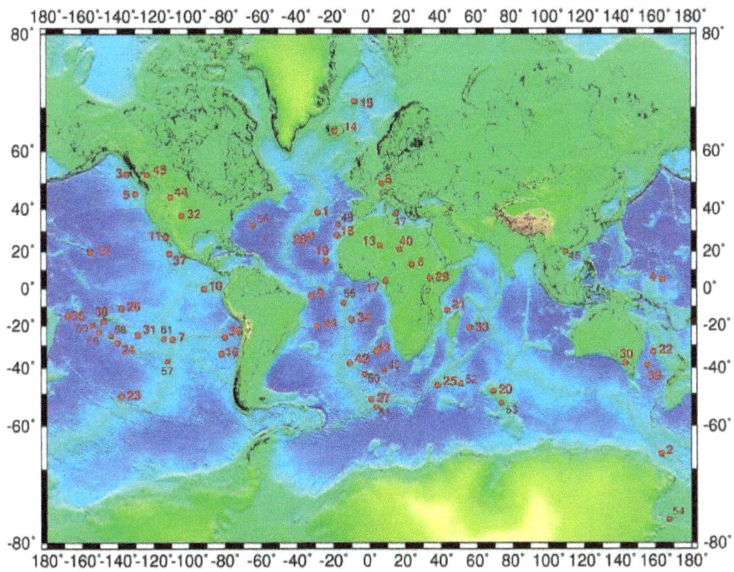

Distribution of hotspots in the list to the left, with the numbers corresponding to those in the list. The Afar hotspot is misplaced.

Eurasian Plate

- Eifel hotspot (8)
  - 50°12′N 6°42′E, w= 1 az= 082° ±8° rate= 12 ±2 mm/yr
- Iceland hotspot (14)
  - 64°24′N 17°18′W
    - Eurasian Plate, w= .8 az= 075° ±10° rate= 5 ±3 mm/yr
    - North American Plate, w= .8 az= 287° ±10° rate= 15 ±5 mm/yr
  - Possibly related to the North Atlantic continental rifting (62 Ma), Greenland.
- Azores hotspot (1)
  - 37°54′N 26°00′W
    - Eurasian Plate, w= .5 az= 110° ±12°
    - North American Plate, w= .3 az= 280° ±15°
- Jan Mayen hotspot (15)

   - 71°N 9°W

- Hainan hotspot (46)
   - 20°N 110°E, az= 000° ±15°

African Plate

- Mount Etna (47)
- 37°45.304′N 14°59.715′E
- Hoggar hotspot (13)
- 23°18′N 5°36′E, w= .3 az= 046° ±12°
- Tibesti hotspot (40)
- 20°48′N 17°30′E, w= .2 az= 030° ±15°
- Jebel Marra/Darfur hotspot (6)
- 13°00′N 24°12′E, w= .5 az= 045° ±8°
- Afar hotspot (29, misplaced in map)
- 7°00′N 39°30′E, w= .2 az= 030° ±15° rate= 16 ±8 mm/yr
- Possibly related to the Afar Triple Junction, 30 Ma.
- Cameroon hotspot (17)
- 2°00′N 5°06′E, w= .3 az= 032° ±3° rate= 15 ±5 mm/yr
- Madeira hotspot (48)
- 32°36′N 17°18′W, w= .3 az= 055° ±15° rate= 8 ±3 mm/yr
- Canary hotspot (18)
- 28°12′N 18°00′W, w= 1 az= 094° ±8° rate= 20 ±4 mm/yr
- New England/Great Meteor hotspot (28)
- 29°24′N 29°12′W, w= .8 az= 040° ±10°
- Cape Verde hotspot (19)
- 16°00′N 24°00′W, w= .2 az= 060° ±30°
- St. Helena hotspot (34)
- 16°30′S 9°30′W, w= 1 az= 078° ±5° rate= 20 ±3 mm/yr
- Gough hotspot (49), at 40°19’ S 9°56’ W.
- 40°18′S 10°00′E, w= .8 az= 079° ±5° rate= 18 ±3 mm/yr
- Tristan hotspot (42), at 37°07′ S 12°17′ W.
- 37°12′S 12°18′W
- Vema hotspot (Vema Seamount, 43), at 31°38’ S 8°20’ E.
- 32°06′S 6°18′W
- Related maybe to the Paraná and Etendeka traps (c. 132 Ma) through the

Walvis Ridge.

- Discovery hotspot (50) (Discovery Seamounts)
- 43°00′S 2°42′W, w= 1 az= 068° ±3°
- Bouvet hotspot (51)
- 54°24′S 3°24′E
- Shona/Meteor hotspot (27)
- 51°24′S 1°00′W, w= .3 az= 074° ±6°
- Réunion hotspot (33)
- 21°12′S 55°42′E, w= .8 az= 047° ±10° rate= 40 ±10 mm/yr
- Possibly related to the Deccan Traps (main events: 68.5–66 Ma)
- Comoros hotspot (21)
- 11°30′S 43°18′E, w= .5 az=118 ±10° rate=35 ±10 mm/yr

Antarctic Plate

- Marion hotspot (25)
    - 46°54′S 37°36′E, w= .5 az= 080° ±12°
- Crozet hotspot (52)
    - 46°06′S 50°12′E, w= .8 az= 109° ±10° rate= 25 ±13 mm/yr
    - Possibly related to the Karoo-Ferrar geologic province (183 Ma)
- Kerguelen hotspot (20)
    - 49°36′S 69°00′E, w= .2 az= 050° ±30° rate= 3 ±1 mm/yr
    - Île Saint-Paul and Île Amsterdam could be part of the Kerguelen hotspot trail (St. Paul is probably not another hotspot)
    - Related maybe to the Kerguelen Plateau (130 Ma)
- Heard hotspot (53)
    - 53°06′S 73°30′E, w= .2 az= 030° ±20°
- Balleny hotspot (2)
    - 67°36′S 164°48′E, w= .2 az= 325° ±7°
- Erebus hotspot (54)
    - 77°30′S 167°12′E

South American Plate

- Trindade/Martin Vaz hotspot (41)
  - 20°30′S 28°48′W, w= 1 az= 264° ±5°
- Fernando hotspot (9)
  - 3°48′S 32°24′W, w= 1 az= 266° ±7°
  - Possibly related to the Central Atlantic Magmatic Province (c. 200 Ma)
- Ascension hotspot (55)
  - 7°54′S 14°18′W

North American Plate

- Bermuda hotspot (56)
  - 32°36′N 64°18′W, w= .3 az= 260° ±15°
- Yellowstone hotspot (44)
  - 44°30′N 110°24′W, w= .8 az= 235° ±5° rate= 26 ±5 mm/yr
  - Possibly related to the Columbia River Basalt Group (17–14 Ma).
- Raton hotspot (32)
  - 36°48′N 104°06′W, w= 1 az= 240°±4° rate= 30 ±20 mm/yr
- Anahim hotspot (45)
  - 52°54′0″N 123°44′0″W (Nazko Cone)

Indo-Australian Plate

- Lord Howe hotspot (22)
  - 34°42′S 159°48′E, w= .8 az= 351° ±10°
- Tasmanid hotspot (Gascoyne Seamount, 39)
  - 40°24′S 155°30′E, w= .8 az= 007° ±5° rate= 63 ±5 mm/yr
- East Australia hotspot (30)
  - 40°48′S 146°00′E, w= .3 az= 000° ±15° rate= 65 ±3 mm/yr

Nazca Plate

- Juan Fernández hotspot (16)

- 33°54′S 81°48′W, w= 1 az= 084° ±3° rate= 80 ±20 mm/yr

■ San Felix hotspot (36)

- 26°24′S 80°06′W, w= .3 az= 083° ±8°

■ Easter hotspot (7)

- 26°24′S 106°30′W, w= 1 az= 087° ±3° rate= 95 ±5 mm/yr

■ Galápagos hotspot (10)

- 0°24′S 91°36′W
  - Nazca Plate, w= 1 az= 096° ±5° rate= 55 ±8 mm/yr
  - Cocos Plate, w= .5 az= 045° ±6°
- Possibly related to the Caribbean large igneous province (main events: 95–88 Ma).

Pacific Plate

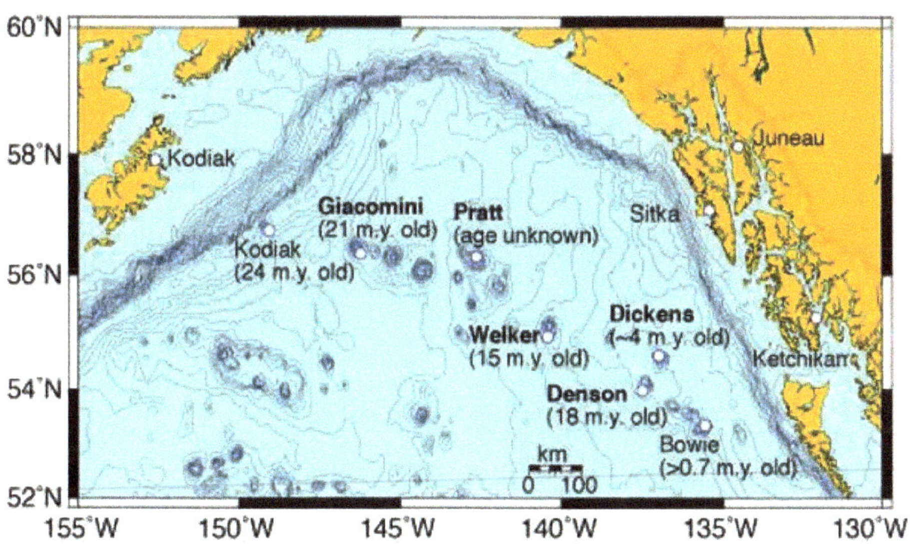

Over millions of years, the Pacific Plate has moved over the Bowie hotspot, creating the Kodiak-Bowie Seamount chain in the Gulf of Alaska.

■ Louisville hotspot (23)

- 53°36′S 140°36′W, w= 1 az= 316° ±5° rate= 67 ±5 mm/yr
- Possibly related to the Ontong Java Plateau (125–120 Ma).

- Foundation hotspot (57)
  - 37°42′S 111°06′W, w= 1 az= 292° ±3° rate= 80 ±6 mm/yr
- Macdonald hotspot (24)
  - 29°00′S 140°18′W, w= 1 az= 289° ±6° rate= 105 ±10 mm/yr
- North Austral/President Thiers (President Thiers Bank, 58)
  - 25°36′S 143°18′W, w= (1.0) azim= 293° ± 3° rate= 75 ±15 mm/yr
- Arago hotspot (Arago Seamount, 59)
  - 23°24′S 150°42′W, w= 1 azim= 296° ±4° rate= 120 ±20 mm/yr
- Maria/Southern Cook hotspot (Îles Maria, 60)
  - 20°12′S 153°48′W, w= 0.8 az= 300° ±4°
- Samoa hotspot (35)
  - 14°30′S 168°12′W, w= .8 az= 285°±5° rate= 95 ±20 mm/yr
- Crough hotspot (Crough Seamount, 61)
  - 26°54′S 114°36′W, w= .8 az= 284° ± 2°
- Pitcairn hotspot (31)
  - 25°24′S 129°18′W, w= 1 az= 293° ±3° rate= 90 ±15 mm/yr
- Society/Tahiti hotspot (38)
  - 18°12′S 148°24′W, w= .8 az= 295°±5° rate= 109 ±10 mm/yr
- Marquesas hotspot (26)
  - 10°30′S 139°00′W, w= .5 az= 319° ±8° rate= 93 ±7 mm/yr
- Caroline hotspot (4)
  - 4°48′N 164°24′E, w= 1 az= 289° ±4° rate= 135 ±20 mm/yr
- Hawaii hotspot (12)
  - 19°00′N 155°12′W, w= 1 az= 304° ±3° rate= 92 ±3 mm/yr
  - Possibly related to the Siberian Traps (251–250 Ma).
- Socorro/Revillagigedos hotspot (37)
  - 19°00′N 111°00′W
- Guadalupe hotspot (11)
  - 27°42′N 114°30′W, w= .8 az= 292° ±5° rate= 80 ±10 mm/yr

- Cobb hotspot (5)

- 46°00′N 130°06′W, w= 1 az= 321° ±5° rate= 43 ±3 mm/yr

- Bowie/Pratt-Welker hotspot (3)
    - 53°00′N 134°48′W, w=.8 az= 306° ±4° rate= 40 ±20 mm/yr

REFERENCES

1. W. J. Morgan (5 March 1971). Convection Plumes in the Lower Mantle. Nature. 230 *(5288): 42–43. Bibcode*:1971Natur.230...42M. doi:10.1038/230042a0.
2. Do plumes exist?. Retrieved 2010-04-25.
3. Foulger, G.R. (2010). Plates vs. Plumes: A Geological Controversy. Wiley-Blackwell. ISBN 978-1-4051-6148-0.
4. Wilson, J. Tuzo (1963). A possible origin of the Hawaiian Islands (PDF). Canadian Journal of Physics. 41 *(6): 863–870. Bibcode*:1963CaJPh..41..863W. doi:10.1139/p63-094.
5. Hotspots: Mantle thermal plumes. United States Geological Survey. 1999-05-05. Retrieved 2008-05-15.
6. Wright, Laura (November 2000). Earth's interior: Raising hot spots. Geotimes. American Geological Institute. Retrieved 2008-06-15.
7. Holbek, Peter (November 1983). Report on Preliminary Geology and Geochemistry of the Ilga Claim Group (PDF). Retrieved 2008-06-15.
8. Axial Seamount. PMEL Earth-Ocean Interactions Program. NOAA. Retrieved 23 September 2014.
9. What the hell is Hawaii?. Retrieved 2011-01-07.
10. Courtillot, V.; Davaillie, A.; Besse, J.; Stock, J. (2003). Three distinct types of hotspots in the Earth's mantle. Earth Sci. Planet. Lett. 205 *(3–4): 295–308. Bibcode*:2003E&PSL.205..295C. doi:10.1016/S0012-821X(02)01048-8.
11. E. V. Verzhbitsky (2003). Geothermal regime and genesis of the Ninety-East and Chagos-Laccadive ridges. Journal of Geodynamics. 35 *(3): 289. Bibcode*:2003JGeo...35..289V. doi:10.1016/S0264-3707(02)00068-6.
12. W. J. Morgan and J. P. Morgan. Plate velocities in hotspot reference frame: electronic supplement (PDF). Retrieved 2011-11-06.
13. Nielsen, Søren B.; Stephenson, Randell; Thomsen, Erik (13 December 2007). Letter:Dynamics of Mid-Palaeocene North Atlantic rifting linked with European intra-plate deformations. Nature. 450 *(7172): 1071–1074. Bibcode*:2007Natur.450.1071N. doi:10.1038/nature06379. PMID 18075591.
14. O'Neill, C.; Müller, R. D.; Steinberger, B. (2003). Revised Indian plate rotations based on the motion of Indian Ocean hotspots(PDF). Earth and Planetary Science Letters. 215: 151–168. Bibcode:2003E&PSL.215..151O. doi:10.1016/S0012-821X(03)00368-6.
15. O'Connor, J. M.; le Roex, A. P. (1992). South Atlantic hot spot-plume systems.

1: Distribution of volcanism in time and space. Earth and Planetary Science Letters. 113 *(3): 343–364. Bibcode*:1992E&PSL.113..343O. doi:10.1016/0012-821X(92)90138-L.

16. Smith, Robert B.; Jordan, Michael; Steinberger, Bernhard; Puskas, Christine M.; Farrell, Jamie; Waite, Gregory P.; Husen, Stephan; Chang, Wu-Lung; O'Connell, Richard (20 November 2009). Geodynamics of the Yellowstone hotspot and mantle plume: Seismic and GPS imaging, kinematics and mantle flow (PDF). Journal of Volcanology and Geothermal Research. 188 *(1–3): 26–56. doi:10.1016/j.jvolgeores.2009.08.020.*
17. Catalogue of Canadian volcanoes- Anahim volcanic belt. Natural Resources Canada. Geological Survey of Canada. Archived from the original on 2011-07-16. Retrieved 2008-06-14.
18. B Steinberger, C Gaina, Plate-tectonic reconstructions predict part of the Hawaiian hotspot track to be preserved in the Bering Sea, Geology 35 (5), 407–410.

CHAPTER 4

Subduction Zone

Chapter Contents

A subduction zone is a convergent boundary where two tectonic plates collide. Plates are large, dense masses in the crust of the Earth, the lithosphere, that float on top of liquefied rock in the asthenosphere. They are constantly shifting and moving, so when they subduct, one pushes beneath the other. These zones create geologic formations such as mountain ranges, ocean trenches, and island arcs, as well as phenomena like earthquakes and volcanoes.

INTRODUCTIUON

Subduction is a geological process that takes place at convergent boundaries of tectonic plates where one plate moves under another and is forced or sinks due to gravity into the mantle. Regions where this process occurs are known as *subduction zones*. Rates of subduction are typically in centimeters per year, with the average rate of convergence being approximately two to eight centimeters per year along most plate boundaries.

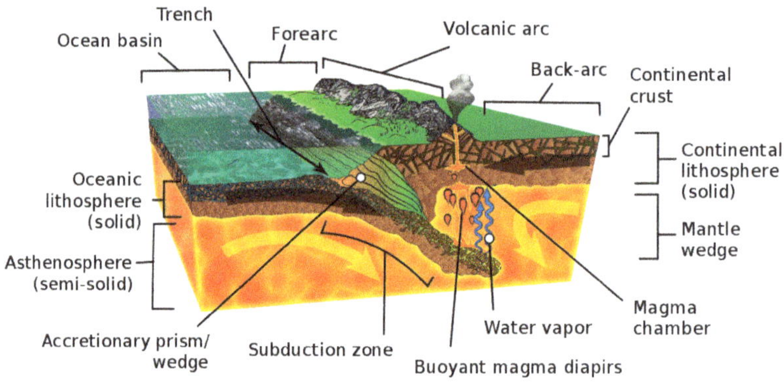

Diagram of the geological process of subduction.

Plates include both oceanic crust and continental crust. Stable subduction zones involve the oceanic lithosphere of one plate sliding beneath the continental or oceanic lithosphere of another plate due to the higher density of the oceanic lithosphere. That is, the subducted lithosphere is always oceanic while the overriding lithosphere may or may not be oceanic. Subduction zones are sites that have a high rate of volcanism, earthquakes, and mountain building.

Orogenesis, or mountain-building, occurs when large pieces of material on the subducting plate (such as island arcs) are pressed into the over-riding plate or when subhorizontal contraction occurs in the over-riding plate. These areas are subject to many earthquakes, which are caused by the interactions between the subducting slab and the mantle, the volcanoes, and (when applicable) the mountain-building related to island arc collisions.

GENERAL DESCRIPTION

Subduction zones are sites of convective downwelling of Earth's lithosphere (the crust plus the top non-convecting portion of the upper mantle). Subduction zones exist at convergent plate boundaries where one plate of oceanic lithosphere converges with another plate. The descending slab, the subducting plate, is over-ridden by the leading edge of the other plate. The slab sinks at an angle of approximately twenty-five to forty-five degrees to Earth's surface. This sinking is driven by the temperature difference between the subducting oceanic lithosphere and the surrounding mantle asthenosphere, as the colder oceanic lithosphere is, on average, denser. At a depth of approximately 80–120 kilometers, the basalt of the oceanic crust is converted to a metamorphic rock called eclogite. At that point, the density of the oceanic crust increases and provides additional negative buoyancy (downwards force). It is at subduction zones that Earth's lithosphere, oceanic crust, sedimentary layers and some trapped water are recycled into the deep mantle.

Earth is so far the only planet where subduction is known to occur. Subduction is the driving force behind plate tectonics, and without it, plate tectonics could not occur.\

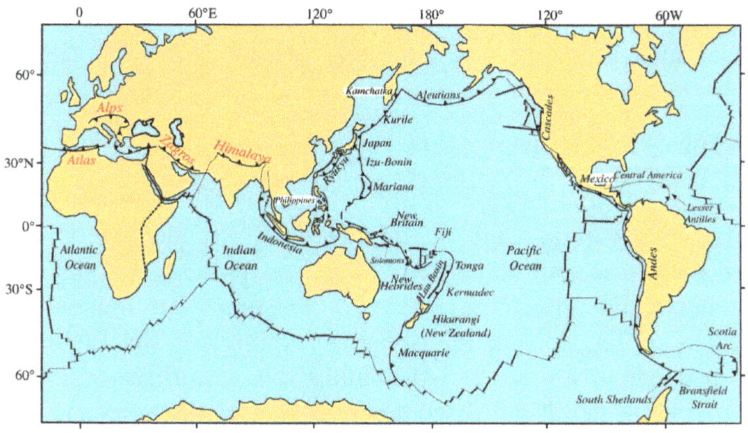

Global convergent plate margins, ~55,000 km long, shown as barbed lines. These are connected to subduction ones at depth. Convergent margins include continental *collision zones* as well as normal subduction zones.

Subduction zones dive down into the mantle beneath 55,000 kilometers of convergent plate margins (Lallemand, 1999), almost equal to the cumulative 60,000 kilometers of mid-ocean ridges. Subduction zones burrow deeply but are imperfectly camouflaged, and geophysics and geochemistry can be used to study them. Not surprisingly, the shallowest portions of subduction zones are known best. Subduction zones are strongly asymmetric for the first several hundred kilometers of their descent. They start to go down at oceanic trenches. Their descents are marked by inclined zones of earthquakes that dip away from the trench beneath the volcanoes and extend down to the 660-kilometre discontinuity. Subduction zones are defined by the inclined array of earthquakes known as the Wadati-Benioff zone after the two scientists who first identified this distinctive aspect. Subduction zone earthquakes occur at greater depths (up to 600 km) than elsewhere on Earth (typically <20 km depth); such deep earthquakes may be driven by deep phase transformations, thermal runaway, or dehydration embrittlement.

The subducting basalt and sediment are normally rich in hydrous minerals and clays. Additionally, large quantities of water are introduced into cracks and fractures created as the subducting slab bends downward. During the transition from basalt to eclogite, these hydrous materials break down, producing copious quantities of water, which at such great pressure and temperature exists as a supercritical fluid. The supercritical water, which is hot and more buoyant than the surrounding rock, rises into the overlying mantle where it lowers the pressure in (and thus the melting temperature of) the mantle rock to the point of actual melting, generating magma. The magmas, in turn, rise because they are less dense than the rocks of the mantle. The mantle-derived magmas (which are basaltic in composition) can continue to rise, ultimately to Earth's surface, resulting in a volcanic eruption. The chemical composition of the erupting lava depends upon the degree to which the mantle-derived basalt interacts with (melts) Earth's crust and/or undergoes fractional crystallization.

Above subduction zones, volcanoes exist in long chains called volcanic arcs. Volcanoes that exist along arcs tend to produce dangerous eruptions because they are rich in water (from the slab and sediments) and tend to be extremely explosive. Krakatoa, Nevado del Ruiz, and Mount Vesuvius are all examples of arc volcanoes. Arcs are also known to be associated with precious metals such as gold, silver and copper believed to be carried by water and concentrated in and around their host volcanoes in rock called ore.

THEORY ON ORIGIN

Although the process of subduction as it occurs today is fairly well understood, its origin remains a matter of discussion and continuing study. Subduction initiation can occur *spontaneously* if denser oceanic lithosphere is able to founder and sink beneath adjacent oceanic or continental lithosphere; alternatively, existing plate motions can *induce* new subduction zones by forcing oceanic lithosphere to rupture and sink into the asthenosphere. Both models can eventually yield self-sustaining subduction zones, as oceanic crust is metamorphosed at great depth and becomes denser than the surrounding mantle rocks. Results from numerical models generally favor induced subduction initiation for most modern subduction zones, which is supported by geologic studies, but other analogue modeling shows the possibility of spontaneous subduction from inherent density differences between two plates at passive margins, and observations from the Izu-Bonin-Mariana subduction system are compatible with spontaneous subduction nucleation. Furthermore, subduction is likely to have spontaneously initiated at some point in Earth's history, as induced subduction nucleation requires existing plate motions, though an unorthodox proposal by A. Yin suggests that meteorite impacts may have contributed to subduction initiation on early Earth.

Geophysicist Don L. Anderson has hypothesized that plate tectonics could not happen without the calcium carbonate laid down by bioforms at the edges of subduction zones. The massive weight of these sediments could be softening the underlying rocks, making them pliable enough to plunge. However, considering that some refractory minerals used for dating early Earth, such as zircon, are typically generated in subduction zones and associated with granites and pegmatites, some of these early dates may have preceded significant biological activity on Earth.

EFFECTS

Metamorphism

A subduction zone is a region of the earth's crust where one tectonic plate moves under another tectonic plate; oceanic crust gets recycled back into the mantle and continental crust gets created by the formation of arc magmas. Arc magmas account for more than 20% of terrestrially produced magmas and are produced by the dehydration of minerals within the subducting slab as it descends into the mantle and are accreted onto the base of the overriding continental plate. Subduction zones host a unique variety of rock types created by the high-pressure, low-temperature conditions a subducting slab encounters during its descent. The metamorphic conditions the slab passes through in this process creates and destroys water bearing (hydrous) mineral phases, releasing water into the mantle. This water lowers the melting point of mantle rock, initiating melting. Understanding the timing and conditions in which these dehydration reactions occur, is key to interpreting mantle melting, volcanic arc magmatism, and the formation of continental crust.

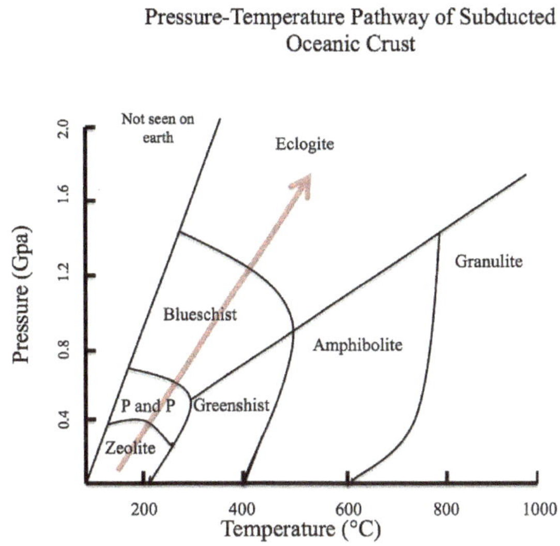

Pressure-temperature pathway for subducted crust.

A metamorphic facies is characterized by a stable mineral assemblage specific to a pressure-temperature range and specific starting material. Subduction zone metamorphism is characterized by a low temperature, high-ultrahigh pressure metamorphic path through the zeolite, prehnite-pumpellyite, blueschist, and eclogite facies stability zones of subducted oceanic crust. Zeolite and prehnite-pumpellyite facies assemblages may or may not be present, thus the onset of metamorphism may only be marked by blueschist facies conditions.Subducting slabs are composed of basaltic crust topped with pelagic sediments; however, the pelagic sediments may be accreted onto the forearc-hanging wall and not subducted. Most metamorphic phase transitions that occur within the subducting slab are prompted by the dehydration of hydrous mineral phases. The breakdown of hydrous mineral phases typically occurs at depths greater than 10 km. Each of these metamorphic facies is marked by the presence of a specific stable mineral assemblage, recording the metamorphic conditions undergone but the subducting slab. Transitions between facies causes hydrous minerals to dehydrate at certain pressure-temperature conditions and can therefore be tracked to melting events in the mantle beneath a volcanic arc.

Oceanic Crust

Arc magmas are produced by the dehydration of minerals contained in oceanic crust formed at mid-ocean ridges. Subducted oceanic crust consists of a four major units. The topmost unit is a thin cap of pelagic sediments up to 0.3 km thick composed of siliceous and calcareous shells, meteoric dusts, and variable amounts of volcanic ash. The next unit is composed of 0.3-0.7 km thick pillow basalts, formed by the quenching of basaltic magma as it erupts into ocean water. Under the pillow basalts is a basaltic sheeted dike complex, that represent cooled magma conduits. The bottom units represent the crystallized magma chamber, feeding the mid-ocean ridge at which the crust was formed. It is composed of

1–5 km thick layered gabbro atop <7 km thick layer of ultramafic rocks (e.g. wehrlite, harzburgite, dunite, and chromite). Oceanic crust is referred to as a metabasite.

Hydrous Minerals of a Subducting Slab

Every year, 1-2 x 10 trillion kilograms of water descend into subduction zones. Approximately 90-95% of that water is contained in hydrous minerals, including mica, phengite, amphibole, lawsonite, chlorite, talc, zoisite, and serpentine. The most significant hydrous minerals are lawsonite (11 wt.% H_2O), phlogopite (2 wt% H_20) and amphibole (2 wt.% H_2O). Phlogopite does not release water until approximately 200 km depth whereas amphibole releases water at approximately 75 km depth. Serpentine is also an important hydrous phase (13 wt.% H2O) that is only present in oceanic crust created at a slow spreading ridge where ultramafic rocks are emplaced at shallow levels. Lawsonite does not release water until approximately 300 km depth and is the last hydrous mineral to do so. Metamorphic dehydration reactions within the slab of trace element hosting minerals such as phengite, lawsonite and, zoisite create a unique type of magma. Arc magmas and the continental crust formed from arc magmas are enriched in boron, lead, arsenic, and antimony derived from the dehydration within the subducting slab. Hydrothermal fluids released from the slab mobilize these elements and allow them to be incorporated into arc magmas, distinguishing arc magmas from those produced at mid-ocean ridges and hotspots.

Facies Transitions and Dehydration Reactions of a Subducting Slab

Zeolite Facies

Basalts may first metamorphose under zeolite facies conditions (50-150 °C and 1–5 km depth) during subduction. Zeolites are microporous silicate minerals that can be produced by the reaction of pore fluids with basalt and pelagic sediments. The zeolite facies conditions typically only affect pelitic sediments undergoing burial, but is commonly displayed by the production of zeolite minerals within the vesicles of vesicular basalt. The glassy rinds on pillow basalts are also susceptible to metamorphism under zeolite facies conditions, which produces the zeolites heulandite or stilbite and hydrous phyllosilicates such as celadonite, smectite, kaolinite, or montmorillonite plus secondary quartz. Crystalline igneous rocks of the subducting slab, such as gabbro and basaltic sheeted dikes, remain stable until greater depth, when the sodium endmember of plagioclase feldspar, albite, replaces detrital igneous plagioclase feldspar. Also at greater depth in thc zeolite facies, the zeolite laumontite replaces the zeolite heulandite and the phyllosilicate chlorite is common.

Prehnite-Pumpellyite Facies

At paths up to 220-320 °C and below 4.5 kbars, subducting slabs may encounter the prehnite-pumpellyite facies, characterized by the presence of the hydrous chlorite, prehnite, albite, pumpellyite, tremolite, and epidote and the loss of the zeolites heulandite and laumonite. Actinolite may occur at higher grade. Aside from albite, these characteristic minerals are water bearing, and may contribute to mantle melting. These minerals are also vital in the formation of glaucophane, which is associated with blueschist facies. The onset of a low-pressure phase of lawsonite is the most significant marker of prehnite-pumpellyite facies metamorphism. The occurrence of lawsonite is significant because lawsonite contains 11 wt.% H_2Owhich is released at higher grade and can initiate significant melting.

Laumontite = Lawsonite + Quartz + H_2O

Blueschist Facies

Blueschist containing the sodic blue amphibole, glaucophane.

Blueschist facies is characterized by the formation of a sodic, blue amphibole, namely, glaucophane, for which the blueschist facies is named. Lawsonite is also diagnostic of blueschist facies and occurs in association with glaucophane. Glaucophane forming reactions are listed below. Glaucophane producing reactions are significant because they can either release water or produce the hydrous phase, lawsonite through the breakdown of hydrous phyllosilicates. At high blueschist facies pressures, albite may break down to from jadeite and quartz. Calcite will commonly pseudomorphose into aragonite under blueschist conditions. Other common minerals of blueschist facies metabasites are paragonite, chlorite, titanite, stilpnomelane, quartz, albite, sericite, and pumpellyite.

Tremolite + Chlorite + Albite = Glaucophane + Epidote + H_2O

Tremolite + Chlorite + Albite = Glaucophane + Lawsonite

Pumpellyite + Chlorite + Albite = Glaucophane + Epidote + H_2O

Eclogite Facies

Transition from blueschist to eclogite facies rock, containing glaucophane, omphacitic pyroxene, and garnet.

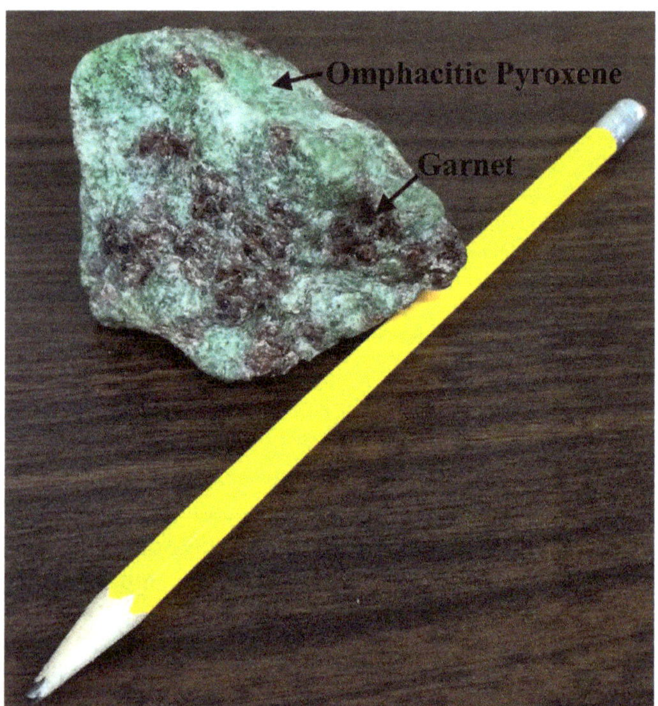

Eclogite facies rock, containing omphacitic pyroxene and garnet.

Eclogite facies is typically encountered around 80–100 km depth and is characterized by the presence of green omphacitic pyroxene and red pyrope garnet. Omphacitic pyroxene is an augite-jadeite solution. At Eclogite facies conditions, plagioclase is no longer stable. The albite component breaks down during glaucophane producing reactions and its sodium becomes incorporated into glaucophane and pyroxene. This reaction is written below. The break down of glaucophane is an important water producing reaction at about 600 °C, and over 1 GPa that can trigger significant mantle melting and volcanism.

Glaucophane + Paragonite = Pyrope + Jadeite + Quartz + H_2O

Another important water producing reaction that occurs during the eclogite facies is the dehydration of the hydrous phyllosilicate phlogopite by the reaction below. This reaction can also trigger significant mantle melting and volcanism. Aside from triggering mantle melt, this reaction may also trigger partial melting of the subducting slab itself.

Phlogopite + Diopside + Orthopyroxene = H_2O + Melt

Lawsonite remains stable up to 1080 °C and 9.4 GPa. the breakdown of lawsonite releases massive amounts of H2O into the mantle that can trigger partial melting of the slab and of the overlying mantle. The breakdown reaction of lawsonite is listed below.

Lawsonite = Grossular + Topaz + Stishovite + H_2O

Antigorite Serpentine is another important water bearing phase that breaks down at eclogite facies conditions. Antigorite breaks down at 600-700 °C and between 2-5 GPa. Antigorite contains 13 wt.% water and therefore causes substantial mantle melting. The reaction is listed below.

Antigorite = Forsterite + Enstatite + H_2O

Transition into the eclogite facies is proposed to be the source of earthquakes at depths greater than 70 km. These earthquakes are caused by the contraction of the slab as minerals transition into more compact crystal structures. The depth of these earthquakes on the subducting slab is known as the Wadati-Benioff Zone

Paired Metamorphic Belts

Paired metamorphic belts are a set of parallel metamorphic rock units parallel to a subduction zone displaying two contrasting metamorphic conditions and thus two distinctive mineral assemblages. Nearest the trench is a zone of low temperature, high pressure metamorphic conditions characterized by blueschist to eclogite facies assemblages. This assemblage is associated with subduction along the trench and low heat flow. Nearest the arc is a zone of high temperature-low pressure metamorphic conditions characterized by amphibolite to granulite facies mineral assemblages such as aluminosilicates, cordierite, and orthopyroxenes. This assemblage is associated with high heat flow generated by melting beneath the volcanic arc.

Volcanic Activity

Volcanoes that occur above subduction zones, such as Mount St. Helens, Mount Etna and Mount Fuji, lie at approximately one hundred kilometres from the trench in arcuate chains, hence the term volcanic arc. Two kinds of arcs are generally observed on Earth: island arcs that form on oceanic lithosphere (for example, the Mariana and the Tonga island arcs), and continental arcs such as the Cascade Volcanic Arc, that form along the coast of continents. Island arcs are produced by the subduction of oceanic lithosphere beneath another oceanic lithosphere (ocean-ocean subduction) while continental arcs formed during subduction of oceanic lithosphere beneath a continental lithosphere (ocean-continent subduction). An example of a volcanic arc having both island and continental arc sections is found behind the Aleutian Trench subduction zone in Alaska.

The arc magmatism occurs one hundred to two hundred kilometres from the trench and approximately one hundred kilometres above the subducting slab. This depth of arc magma generation is the consequence of the interaction between hydrous fluids, released from the subducting slab, and the arc mantle wedge that is hot enough to melt with the addition of water. Arcs produce about 25% of the total volume of magma produced each year on Earth (approximately thirty to thirty-five cubic kilometres), much less than the volume produced at mid-ocean ridges, and they contribute to the formation of new continental crust. Arc volcanism has the greatest impact on humans, because many arc volcanoes lie above sea level and erupt violently. Aerosols injected into the stratosphere during violent eruptions can cause rapid cooling of Earth's climate and affect air travel.

Earthquakes and Tsunamis

The strains caused by plate convergence in subduction zones cause at least three different types of earthquakes. Earthquakes mainly propagate in the cold subducting slab and define the Wadati-Benioff zone. Seismicity shows that the slab can be tracked down to the upper mantle/lower mantle boundary (approximately six hundred kilometer depth).

Nine of the ten largest earthquakes of the last 100 years were subduction zone events, which included the 1960 Great Chilean earthquake, which, at M 9.5, was the largest earthquake ever recorded; the 2004 Indian Ocean earthquake and tsunami; and the 2011 Tōhoku earthquake and tsunami. The subduction of cold oceanic crust into the mantle depresses the local geothermal gradient and causes a larger portion of Earth to deform in a more brittle fashion than it would in a normal geothermal gradient setting. Because earthquakes can occur only when a rock is deforming in a brittle fashion, subduction zones can cause large earthquakes. If such a quake causes rapid deformation of the sea floor, there is potential for tsunamis, such as the earthquake caused by subduction of the Indo-Australian Plate under the Euro-Asian Plate on December 26, 2004 that devastated the areas around the Indian Ocean. Small tremors which cause small, nondamaging tsunamis, also occur frequently.

Outer rise earthquakes occur when normal faults oceanward of the subduction zone are activated by flexture of the plate as it bends into the subduction zone. The Samoa earthquake of 2009 is an example of this type of event. Displacement of the sea floor caused by this event generated a six-metre tsunami in nearby Samoa.

Anomalously deep events are a characteristic of subduction zones, which produce the deepest quakes on the planet. Earthquakes are generally restricted to the shallow, brittle parts of the crust, generally at depths of less than twenty kilometres. However, in subduction zones, quakes occur at depths as great as seven hundred kilometres. These quakes define inclined zones of seismicity known as Wadati-Benioff zones, after the scientists who discovered them, which trace the descending lithosphere. Seismic tomography has helped detect subducted lithosphere in regions where there are no earthquakes. Some subducted slabs seem not to be able to penetrate the major discontinuity in the mantle that lies at a depth of about 670 kilometres whereas other subducted oceanic plates can penetrate all the way to the core-mantle boundary. The great seismic discontinuities in the mantle, at 410 and 670 kilometre depth, are disrupted by the descent of cold slabs in deep subduction zones.

Orogeny

Orogeny is the process of mountain building. Subducting plates can lead to orogeny by bringing oceanic islands, oceanic plateaus, and sediments to convergent margins. The material often does not subduct with the rest of the plate but instead is accreted (scraped off) to the continent resulting in exotic terranes. The collision of this oceanic material causes crustal thickening and mountain-building. This accretion process is thought by many geologists to be the reason for the crustal growth of western North America and of the uplift that produced the Rocky Mountains.

Subduction may also cause orogeny without bringing in oceanic material that collides with the overriding continent. When the subducting plate subducts at a shallow angle underneath a continent (something called "flat-slab subduction"), the subducting plate may have enough traction on the bottom of the continental plate to cause the upper plate to contract leading to folding, faulting, crustal thickening and mountain building. This flat-slab subduction process is thought to be one of the main causes of mountain building and deformation in South America.

The processes described above allow subduction to continue while mountain building happens progressively, which is in contrast to continent-continent collision orogeny, which often leads to the termination of subduction.

SUBDUCTION ANGLE

Subduction typically occurs at a moderately steep angle right at the point of the convergent plate boundary. However, anomalous shallower angles of subduction are known to exist as well some that are extremely steep.

- **Flat-slab subduction (<30°):** occurs when subducting lithosphere, called a slab, subducts horizontally or nearly horizontally. The flat slab can extend for hundreds of kilometres and can even extend to over a thousand. That is abnormal, as the dense slab typically sinks at a much steeper angle directly at the subduction zone. Because subduction of slabs to depth is necessary to drive subduction zone volcanism (through the destabilization and dewatering of minerals and the resultant flux melting of the mantle wedge), flat-slab subduction can be invoked to explain volcanic gaps. Flat-slab subduction is ongoing beneath part of the Andes causing segmentation of the Andean Volcanic Belt into four zones. The flat-slab subduction in northern Peru and Norte Chico region of Chile is believed to be the result of the subduction of two buoyant aseismic ridges, the Nazca Ridge and the Juan Fernández Ridge respectively. Around Taitao Peninsula flat-slab subduction is attributed to the subduction of the Chile Rise, a spreading ridge. The Laramide Orogeny in the Rocky Mountains of United States is attributed to flat-slab subduction. Then, a broad volcanic gap appeared at the southwestern margin of North America, and deformation occurred much farther inland; it was during this time that the basement-cored mountain ranges of Colorado, Utah, Wyoming, South Dakota, and New Mexico came into being. The most massive subduction zone earthquakes, so-called megaquakes, have been found to occur in flat-slab subduction zones.

- **Steep-angle subduction (>70°):** occurs in subduction zones where Earth's oceanic crust and lithosphere are old and thick and have, therefore, lost buoyancy. The steepest dipping subduction zone lies in the Mariana Trench, which is also where the oceanic crust, of Jurassic age, is the oldest on Earth exempting ophiolites. Steep-angle subduction is, in contrast to flat-slab subduction, associated with back-arc extension of crust making volcanic arcs and fragments of continental crust wander away from continents over geological times leaving behind a marginal sea.

IMPORTANCE

Subduction zones are important for several reasons:

- Subduction Zone Physics: Sinking of the oceanic lithosphere (sediments, crust, mantle), by contrast of density between the cold and old lithosphere and the hot asthenospheric mantle wedge, is the strongest force (but not the only one) needed to drive plate motion and is the dominant mode of mantle convection.
- Subduction Zone Chemistry: The subducted sediments and crust dehydrate and release water-rich (aqueous) fluids into the overlying mantle, causing mantle melting and fractionation of elements between surface and deep mantle reservoirs, producing island arcs and continental crust.
- Subduction zones drag down subducted oceanic sediments, oceanic crust, and mantle lithosphere that interact with the hot asthenospheric mantle from the overriding plate to produce calc-alkaline series melts, ore deposits, and continental crust.

Subduction zones have also been considered as possible disposal sites for nuclear waste in which the action of subduction itself would carry the material into the planetary mantle, safely away from any possible influence on humanity or the surface environment. However, that method of disposal is currently banned by international agreement. Furthermore, plate subduction zones are associated with very large megathrust earthquakes, making the effects on using any specific site for disposal unpredictable and possibly adverse to the safety of longterm disposal.

FLAT SLAB SUBDUCTION

Flat slab subduction is characterized by a low subduction angle (<30 degrees to horizontal) beyond the seismogenic zone and a resumption of normal subduction far from the trench. A slab is slang for the subducting lower plate. Although, some would characterize flat slab subduction as any shallowly dipping lower plate as in western Mexico. Flat slab subduction is associated with the pinching out of the asthenosphere, an inland migration of arc magmatism (magmatic sweep), and an eventual cessation of arc magmatism. The coupling of the flat slab to the upper plate is thought to change the style of deformation occurring on the upper plate's surface and form basement-cored uplifts like the Rocky Mountains. The flat slab also may hydrate the lower continental lithosphere and be involved in the formation of economically important ore deposits. The failure of a flat slab is associated with ignimbritic volcanism and the reverse migration of arc volcanism. Multiple working hypotheses about the cause of flat slabs are subduction of thick, buoyant oceanic crust (15–20 km) and trench rollback accompanying a rapidly overriding upper plate and enhanced trench suction. The west coast of South America has two of the largest flat slab subduction zones. Flat slab subduction is occurring at 10% of subduction zones.

The idea has its beginnings in the late seventies. Seismic studies of the Andean margin seemed to show a zone of subhorizontal lower plate at a depth of 100 km. The Cornell-Carnigie debate between Cornell University geophysicists and workers at the Carnegie Institute of Washington centered on whether local deployments of seismometers would yield better results than looking at global (teleseismic) data. The Carnegie Institution seemed to have won the day with the local deployment imaging the flat slab where teleseismic data argued for a shallowing dipping slab with no near horizontal zone.The idea was taken up to explain the Laramide orogeny, as the flat slab subduction zones on the Andean margin are associated with more inboard surface deformation and magmatic gaps. Flat slab subduction is an active area of research; the causal mechanisms for its occurrence have not been sorted out.

Causal Mechanisms & Consequences of Flat Slab Subduction

Causal Mechanisms

There are several working hypotheses for the initiation of flat slab subduction. The buoyant ridge hypothesis seems to favored at the moment.

Subduction of Buoyant Oceanic Crust

The subduction of bathymetric highs such as aseismic ridges, oceanic plateaus, and seamounts has been posited as the primary driver of flat slab subduction. The Andean flat slab subduction zones, the Peruvian slab and the Pampean (Chilean) flat slab, are spatially correlated with the subduction of bathymetric highs, the Nazca Ridge and the Juan Fernandéz Ridge, respectively. The thick, buoyant oceanic crust lowers the density of the slab, and the slab fails to sink into the mantle after coming to a shallow depth

(~100 km) due the lessened density contrast. This is supported by the fact that all slabs under ~50 Ma. However, there are cases where aseismic ridges on the same scale as the Nazca Ridge are subducting normally, and cases where flat slabs are not associated with bathymetric highs. There are few flat slabs in the Western Pacific in areas associated with the subduction of bathymetric highs. Geodynamic modeling has called into question whether buoyant oceanic crust alone can generate flat slab subduction.

Trenchward Motion of Overriding Plate with Cratonic Keel

Another explanation for slab flattening is the lateral movement of the overriding plate in a direction opposite to that of the downgoing slab. The overriding plate is often equipped with a cratonic keel of thick continental lithosphere which, if close enough to the trench, can impinge upon the flow in the mantle wedge. Trench suction is included in this causal mechanism. Trench suction is induced by the flow of the asthenosphere in the mantle wedge area; trench suction increases with subduction velocity, a decrease of the mantle wedge thickness, or an increase in the mantle wedge viscosity. Trench retreat is motion of the trench in a direction opposite to that of plate convergence thought to be related to the position of the trench along the larger subduction zone with retreat occurring near the edges of subduction zones. Modeling experiments have shown that if the cratonic lithosphere is thick and the trench retreats, the shutdown of the mantle wedge increases trench suction to an extent that the slab flattens.

Consequences

Delay in the Eclogitization

Eclogite is a dense (3.5 g/cu. cm), garnet-bearing rock that is formed as the oceanic crust subducts to zones of high pressure and temperature. The reaction that forms eclogite dehydrates the slab, and hydrates the mantle wedge above. The now denser slab more effectively sinks. A delay in eclogitization could arise through the subduction of zone thicker oceanic lithosphere without deeply penetrating faults. Oceanic crust is normally faulted at the trench rise by the bending of the plate as it subducts. This may be an effect or a cause of flat slab subduction, but it seems as though it is more likely an effect. A resumption of normally dipping subduction beyond the flat slab portion is associated with the eclogite reaction, and the amount of time needed to accumulate enough eclogite for the slab to start sinking may be what limits temporal scale of flat slab subduction.

Magmatic Gaps and Adakitic Volcanism

As the subducting plate flattens there is an inboard migration in the magmatic arc that can be tracked. In the Chilean flat slab region (~31-32 degrees S), around 7-5 Ma there was an eastward migration, broadening and gradual shutdown down of the volcanic arc associated with slab flattening. This occurs as the previous magmatic arc position on the upper plate (100–150 km above subducting plate) is no longer aligned with the zone of partial melting above the flattening slab. The magmatic arc migrates to a new location that coincides with the zone of partial melting above the flattening slab. Magmatism before the Laramide orogeny migrated all the way to western South Dakota. Eventually, the magmatic activity above the flat slab may completely cease as the subducting plate and upper plate pinch out the mantle wedge. Upon the failure of the flat slab, the mantle wedge can again start circulating hot asthenosphere (1300 degrees C) in an area that has been heavily hydrated, but that had not produced any melt; this leads to widespread ignimbritic volcanism, which is seen in both the Andean flat slab effected regions and the western United States.

Adakites are dacitic and andesitic magmas that are highly depleted in heavy rare earth elements and high strontium/yttrium ratios and may be derived of melting of the oceanic crust. Adakites are thought to erupt or be emplaced during the transition from normally dipping subduction to flat subduction as the magmatic arc widens and migrates more inland. Adakitic rocks can be seen in modern Ecuador, a possible incipient flat slab zone, and in central Chile there are 10-5 Ma adakitic rocks.Thus, adakitic rocks could be used as marker of past episodes of flat slab subduction.

Surface Deformation

Flat slabs are thought to result in zones of broad, diffuse deformation in the upper plate located far landward from the trench.Flat slab subduction is associated with basement-cored uplifts also known as "thick-skinned" deformation of the overriding plate like the Sierra Pampeanas in South America possibly associated with the subduction of the Juan Fernandéz Ridge.These areas of basement-cored uplifts are visually correlated with flat slab subduction zones. In contrast, thin-skinned deformation is the normal mode of upper plate deformation, and does not involve basement rock. Crustal shortening is observed to extend farther inland than in normally dipping subduction zones; the Sierra Pampeanas are over 650 km east of the trench axis. Flat slabs have been used as an explanation for the Laramide Orogeny and the central Altiplano-Puna region. Another interesting feature that may be associated with the flat slab subduction of the Nazca Ridge is the Fitzcarrald arch located in the Amazonian Basin. The Fitzcarrald arch is a long-wavelength, linear topographic feature extending from eastern Peru to western Brazil beyond the Subandean thrust front into an undeformed area and rising ~600 masl. The Fitzcarrald arch has the effect of splitting the Amazonian Basin into three subbasins: northern Amazonian foreland basin, southern Amazonian foreland basin, and the eastern Amazonian foreland basin.

Seismicity

The shape of the flat slab is constrained through earthquakes within the subducting slab and the interface between the upper plate and the subducting slab. Flat slab zones along the Andean margin release 3-5 times more energy through upper plate earthquakes than adjacent, more steeply dipping subduction zones. Upper plate earthquake focal mechanisms indicate that stress is aligned parallel with motion of the plate, and that stress is transmitted high into the upper plate from the lower. The reason for this enhanced seismicity is more effective coupling of the upper and lower plates. In normal subduction zones the coupling interface, the area in which the two plates are in close proximity, between the two plates is ~100–200 km long, but in flat slab subduction zones the coupling interface is much longer, 400–500 km. Although the lower lithosphere of the upper deforms plastically, numerical modeling has shown stress can be transmitted to crustal regions which behave in a brittle fashion. Along the subducting plate seismicity is more variable, especially intermediate-depth earthquakes. The variability may be controlled by the thickness of the crust and how efficiently it can release water. Thick crust that is not as deeply fractured by trench rise normal faulting may not dehydrate rapidly enough to induce intermediate-depth earthquakes. The Peruvian flat slab lacks significant intermediate-depth earthquakes and is associated with the subduction of the ~17 km thick Nazca Ridge.

Andean Flat Slabs

In the late seventies early research recognized the unique nature of the two large flat slab subduction zones along the Andean margin of South America. Two large and one smaller current flat slab subduction segments exist along the Andean margin: the Peruvian, Pampean (Chilean), and the Bucaramanga. Three Cenozoic flat slab segment are also known: Altiplano, Puna, and Payenia.

The Peruvian flat slab is located between the Gulf of Guayaquil (5 degrees S) and Arequipa (14 degrees S), extending ~1500 km along the strike of the subduction zone. The Peruvian flat slab is the largest in the world, and extends ~700 km inboard from the trench axis. The subducting plate starts at a dip of 30 degrees then flattens out at a depth of 100 km under the Eastern Cordillera and Subandean zone. The segment is visually correlated with the subduction of the Nazca Ridge, an aseismic ridge with thickened crust. The second highest zone in the Andes, Cordillera Blanca, is associated with the Peruvian flat slab segment and uplift of basement-cored blocks. Volcanism in the area ceased in the Late Miocene (11-5 Ma). Plate reconstructions time the collision of the Nazca Ridge with the subduction zone at 11.2 Ma at 11 degree S, which implies that the northern extent of the Peruvian flat slab may require some other subducted feature like an oceanic plateau. A putative subducted plateau, the Inca Plateau, has been argued for.

The Pampean or Chilean flat slab segment is located between 27 degrees S and 33 degrees S, extending ~550 km along the strike of the subduction zone. The Pampean flat slab similarly extends ~700 km inboard from the trench axis. The segment is visually correlated with the Juan Fernandez Ridge, and the highest peak in the Andes, the non-volcanic Aconcagua (6961 m). This area has undergone the same thick-skinned deformation, leading to the high mountain peaks.

The Bucaramanga segment was recognized in early eighties from limited seismological evidence. The segment is encompassed between 6 and 9 degrees N in Colombia, extending ~350 km along the strike of the subduction zone.

Other Flat Slabs

There are several other flat slab segments that warrant a mention:

- Alaskan: 145-150 degrees W along the Aleutian Trench associated with the Yakutat microplate
- Costa Rica: 82-84 degrees W associated with the Cocos Ridge

- Mexico: 96-100 degrees W associated with the Tenhuantepec Ridge
- Cascadian, United States: 46-49 degrees N associated with the subduction of young oceanic crust

Economic Geology

Subduction of thick oceanic crust could be linked with the metallogenesis of copper and gold deposits. The 10 largest young (<18 Ma) gold deposits in South America are associated with flat slab segments. Enhanced metallogenesis may be caused by the cessation of magmatism in the arc allowing the conservation of sulfur-rich volatiles. The failure of the putative flat slab under western North America may have been vital in producing Carlin-type gold deposits.

Early Earth Subduction

Early Earth's mantle was hotter and it has been proposed that flat slab subduction was the dominant style. Computer modeling has shown that an increase in oceanic plate buoyancy associated with enhanced oceanic crust production would have the been counteracted by decreased mantle viscosity, so flat slab subduction would not have been dominant or non-existent.

Erosion and Tectonics

The interaction between **erosion and tectonics** has been a topic of debate since the early 1990s. While the tectonic effects on surface processes such as erosion have long been recognized (for example, river formation as a result of tectonic uplift), the opposite (erosional effects on tectonic activity) has only recently been addressed. The primary questions surrounding this topic are what types of interactions exist between erosion and tectonics and what are the implications of these interactions. While this is still a matter of debate, one thing is clear, the Earth's landscape is a product of two factors: tectonics, which can create topography and maintain relief through surface and rock uplift, and climate, which mediates the erosional processes that wear away upland areas over time. The interaction of these processes can form, modify, or destroy geomorphic features on the Earth's surface.

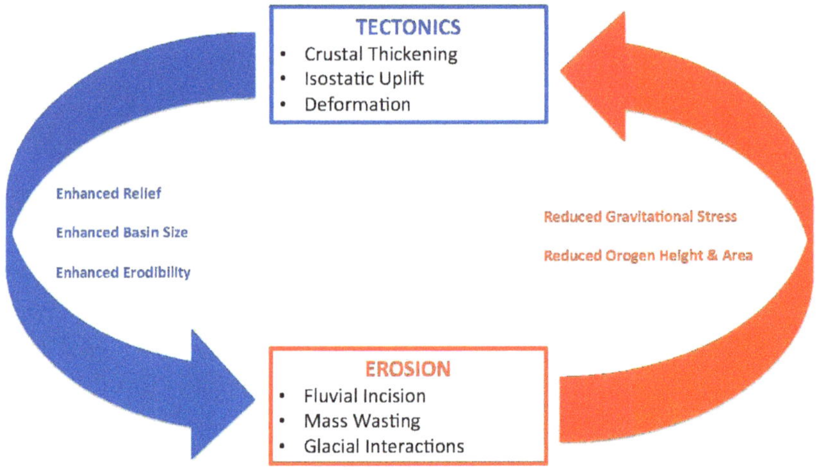

Interactions and feedback pathways for tectonics and erosional processes

Tectonic Processes

The term tectonics refers to the study of the Earth's surface structure and the ways in which it changes over time. Tectonic processes typically occur at plate boundaries which are one of three types: convergent boundaries, divergent boundaries, or transform boundaries. These processes form and modify the topography of the Earth's surface, effectively increasing relief through the mechanisms of isostatic uplift, crustal thickening, and deformation in the form of faulting and folding. Increased elevations, in relation to regional base levels, lead to steeper river channel gradients and an increase in orographically localized precipitation, ultimately resulting in drastically increased erosion rates. The topography, and general relief, of a given area determines the velocity at which surface runoff will flow, ultimately determining the potential erosive power of the runoff. Longer, steeper slopes are more prone to higher rates of erosion during periods of heavy rainfall than shorter, gradually sloping areas. Thus, large mountain ranges, and other areas of high relief, formed through tectonic uplift will have significantly higher rates of erosion. Additionally, tectonics can directly influence erosion rates on a short timescale, as is clear in the case of earthquakes, which can trigger landslides and weaken surrounding rock through seismic disturbances.

While tectonic uplift in any case will lead to some form of increased elevation, thus higher rates of erosion, a primary focus is set on isostatic uplift as it provides a fundamental connection between the causes and effects of erosional-tectonic interactions.

Isostatic Uplift

Understanding the principle of isostasy is a key element to understanding the interactions and feedbacks shared between erosion and tectonics. The principle of isostasy states that when free to move vertically, lithosphere floats at an appropriate level in the asthenosphere so that the pressure at a depth of compensation in the asthenosphere well below the base of the lithosphere is the same. Isostatic uplift is both a cause and an effect of erosion. When deformation occurs in the form of crustal thickening an isostatic response is induced causing the thickened crust to sink, and surrounding thinner crust to uplift. The resulting surface uplift leads to enhanced elevations, which in turn induces erosion. Alternatively, when a large amount of material is eroded away from the Earth's surface uplift occurs in order to maintain isostatic equilibrium. Because of isostasy, high erosion rates over significant horizontal areas can effectively suck up material from the lower crust and/ or upper mantle. This process is known as *isostatic rebound* and is analogous to Earth's response following the removal of large glacial ice sheets.

Isostatic uplift and corresponding erosion are responsible for the formation of regional-scale geologic features as well as localized structures. Two such examples include:

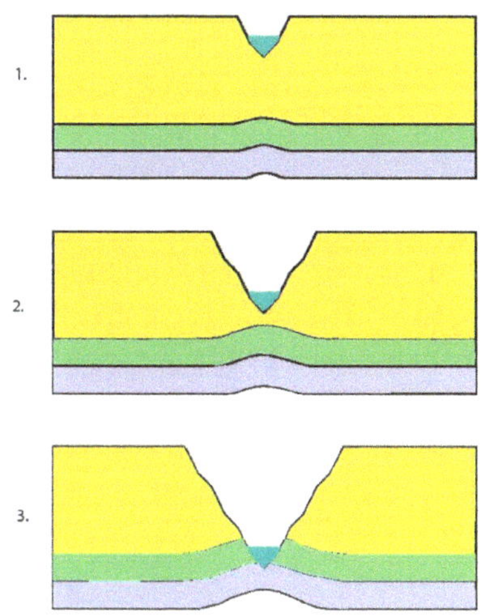

Formation of a river anticline.

- **Continental shields-** Generally large areas of low relief (<100 m) in the Earth's crust where Precambrian crystalline igneous and high-grade metamorphic rocks are exposed. Shields are considered tectonically stable areas in comparison to the activity occurring at their margins and the boundaries between plates, but their formation required large amounts of tectonic activity and erosion. Shields, along with stable platforms, are the basic tectonic components of continents, therefore understanding their development is critical to understanding the development of other surface features on Earth. Initially, a mountain belt is formed at a convergent plate margin. Transformation of a mountain belt to a shield is majorly dependent on two factors: (1) erosion of the mountain belt by running water and (2) isostatic adjustment resulting from the removal of surface rock due to erosion. This process of erosion followed by isostatic adjustment continues until the system is at isostatic equilibrium. At this point large-scale erosion can no longer occur because the surface has eroded down to nearly sea-level and uplift ceases due to the system's state of equilibrium.

- **River anticlines**- Geologic structures formed through the focused uplift of rock underlying confined areas of high erosion (*i.e.*, rivers). Isostatic rebound resulting from the rapid removal of overlying rock, via erosion, causes the weakened areas of crustal rock to uplift from the apex of the river. In order for the development of these structures to occur the erosion rate of the river must exceed both the average erosional rate of the area, and the rate of uplift of the orogen. The two factors influencing the development of these structures are stream power of the associated river and the flexural rigidity of the crust in the area. The combination of increased stream power with decreased flexural rigidity results in the system's progression from a transverse anticline to a river anticline.

Channel Flow

Channel flow describes the process through which hot, viscous crustal material flows horizontally between the upper crust and lithospheric mantle, and is eventually pushed to the surface. This model aims to explain features common to metamorphic hinterlands of some collisional orogens, most notably the Himalaya–Tibetan plateau system. In mountainous areas with heavy rainfall (thus, high erosion rates) deeply incising rivers will form. As these rivers wear away the Earth's surface two things occur: (1) pressure is reduced on the underlying rocks effectively making them weaker and (2) the underlying material moves closer to the surface. This reduction of crustal strength, coupled with the erosional exhumation, allows for the diversion of the underlying channel flow toward the Earth's surface.

Erosional Processes

A natural arch produced by erosion of differentially weathered rock in Jebel Kharaz (Jordan).

The term erosion refers to the group of natural processes, including weathering, dissolution, abrasion, corrosion, and transportation, by which material is worn away from the Earth's surface to be transported and deposited in other locations.

- **Differential erosion**- Erosion that occurs at irregular or varying rates, caused by the differences in the resistance and hardness of surface materials; softer and weaker rocks are rapidly worn away, whereas harder and more resistant rocks remain to form ridges, hills, or mountains. Differential erosion, along with the tectonic setting, are two of the most important controls on the evolution of continental landscapes on Earth.

The feedback of erosion on tectonics is given by the transportation of surface, or near-surface, mass (rock, soil, sand, regolith, etc.) to a new location. This redistribution of material can have profound effects on the state of gravitational stresses in the area, dependent on the magnitude of mass transported. Because tectonic processes are highly dependent on the current state of gravitational stresses, redistribution of surface material can lead to tectonic activity. While erosion in all of its forms, by definition, wears away material from the Earth's surface, the process of mass wasting as a product of deep fluvial incision has the highest tectonic implications.

Mass Wasting

Talus cones produced by mass wasting, north shore of Isfjord, Svalbard, Norway.

Mass wasting is the geomorphic process by which surface material move downslope typically as a mass, largely under the force of gravity As rivers flow down steeply sloping mountains, deep channel incision occurs as the river's flow wears away the underlying rock. Large channel incision progressively decreases the amount of gravitational force needed for a slope failure event to occur, eventually resulting in mass wasting.Removal of large amounts of surface mass in this fashion will induce an isostatic response resulting in uplift until equilibrium is reached.

Effects on Structural Evolution

Recent studies have shown that erosional and tectonic processes have an effect on the structural evolution of some geologic features, most notably orogenic wedges. Highly useful sand box models, in which horizontal layers of sand are slowly pressed against a backstop, have shown that the geometries, structures, and kinematics of orogenic wedge formation with and without erosion and sedimentation are significantly different. Numerical models also show that the evolution of orogens, their final tectonic structure, and the potential development of a high plateau, all are sensitive to the long term climate over the mountains, for example, the concentration of precipitation in one side of the orogen due to orographic lift under a dominant wind direction.

VOLCANIC ARC

A **volcanic arc** is a chain of volcanoes formed above a subducting plate,positioned in an arc shape as seen from above. Offshore volcanoes form islands, resulting in a volcanic island arc. Generally, volcanic arcs result from the subduction of an oceanic tectonic plate under another tectonic plate, and often parallel an oceanic trench. The oceanic plate is saturated with water, and volatiles such as water drastically lower the melting point of the mantle. As the oceanic plate is subducted, it is subjected to greater and greater pressures with increasing depth. This pressure squeezes water out of the plate and introduces it to the mantle. Here the mantle melts and forms magma at depth under the overriding plate. The magma ascends to form an arc of volcanoes parallel to the subduction zone.

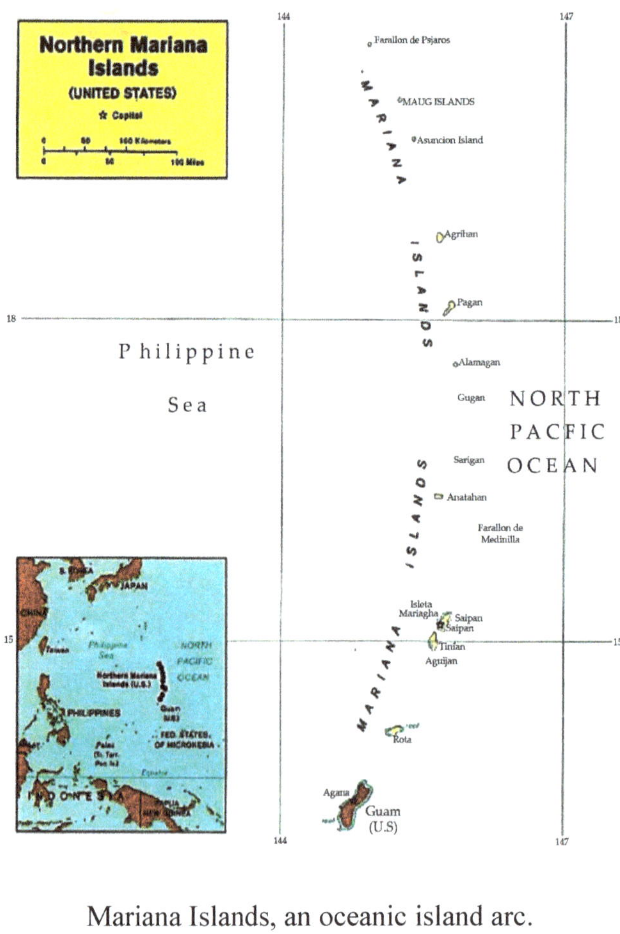

Mariana Islands, an oceanic island arc.

These should not be confused with hotspot volcanic chains, where volcanoes often form one after another in the middle of a tectonic plate, as the plate moves over the hotspot, and so the volcanoes progress in age from one end of the chain to the other. The Hawaiian Islands form a typical hotspot chain; the older islands (tens of millions of years old) to the northwest are smaller and more lush than the recently created (400,000 years ago) Hawaii island itself, which is more rocky. Hotspot volcanoes are also known as "intra-plate" volcanoes, and the islands they create are known as Volcanic Ocean Islands. Volcanic arcs do not generally exhibit such a simple age-pattern.

There are two types of volcanic arcs:

- **oceanic arcs** form when oceanic crust subducts beneath other oceanic crust on an adjacent plate, creating a volcanic island arc. (Not all island arcs are volcanic island arcs.)
- **continental arcs** form when oceanic crust subducts beneath continental crust on an adjacent plate, creating an arc-shaped mountain belt. (Not all mountain belts form in this way.)

In some situations, a single subduction zone may show both aspects along its length, as part of a plate subducts beneath a continent and part beneath adjacent oceanic crust.

(The term volcanic arc is often confused with the term island arc. The former is a series of volcanoes, but not necessarily offshore. The latter is a series of islands, not necessarily composed solely of volcanoes.)

Volcanoes are present in almost any mountain belt, but this does not make it a **volcanic arc**. Often there are isolated, but impressively huge volcanoes in a mountain belt. For instance, Vesuvius and the Etna volcanoes in Italy are part of separate but different kinds of mountainous volcanic ensembles.

The active front of a volcanic arc is the belt where volcanism develops at a given time. Active fronts may move over time (millions of years), changing their distance from the oceanic trench as well as their width.

Petrology

In the subduction zone, loss of water from the subducted slab induces partial melting of the overriding mantle and generates low-density, calc-alkaline magma that buoyantly rises to intrude and be extruded through the lithosphere of the overriding plate. This loss of water is due to the destabilization of the mineral chlorite at approximately 40–60 km depth. This is the reason for island arc volcanism at consistent distances from the subducting slab: because the temperature-pressure conditions for flux-melting volcanism due to chlorite destabilization will always occur at the same depth, the distance from the trench to the arc volcanoes is determined only by the dip angle of the subducting slab.

On the subducting side of the island arc is a deep and narrow oceanic trench, which is the trace at the Earth's surface of the boundary between the downgoing and overriding plates. This trench is created by the gravitational pull of the relatively dense subducting plate pulling the leading edge of the plate downward. Multiple earthquakes occur along this subduction boundary with the seismic hypocenters located at increasing depth under the island arc: these quakes define the Wadati-Benioff zones. The volcanic arc forms when the subducting plate reaches a depth of about 100 km.

Ocean basins that are being reduced by subduction are called ‹remnant oceans' as they will slowly be shrunken out of existence and crushed in the subsequent orogenic collision. This process has happened over and over in the geologic history of the Earth.

In the rock record, volcanic arcs can be seen as the volcanic rocks themselves, but because volcanic rock is easily weathered and eroded, it is more typical that they are seen as plutonic rocks, the rocks that formed underneath the arc (e.g. the Sierra Nevada batholith), or in the sedimentary record as lithic sandstones.

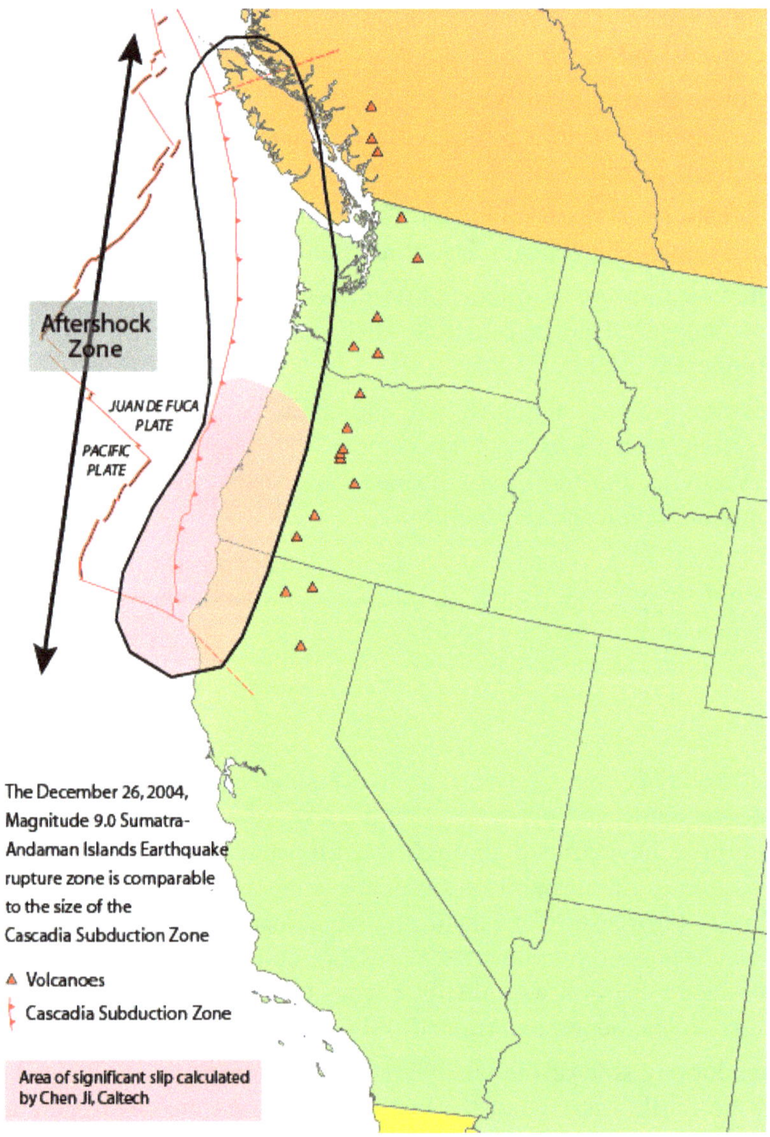

Cascade Volcanic Arc, a continental volcanic arc

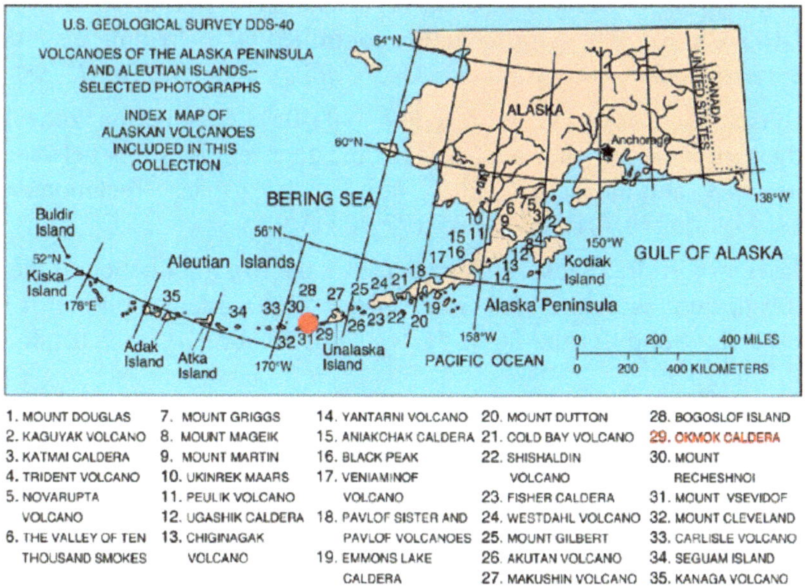

The Aleutian Arc, with both oceanic and continental parts.

REFERENCES

1. Kumar, Abhash; Wagner, Lara S.; Beck, Susan L.; Long, Maureen D.; Zandt, George; Young, Bissett; Tavera, Hernando; Minaya, Estella (2016-05-01). "Seismicity and state of stress in the central and southern Peruvian flat slab". Earth and Planetary Science Letters. 441: 71–80. doi:10.1016/j.epsl.2016.02.023.
2. Humphreys, Eugene; Hessler, Erin; Dueker, Kenneth; Farmer, G. Lang; Erslev, Eric; Atwater, Tanya (2003-07-01). "How Laramide-Age Hydration of North American Lithosphere by the Farallon Slab Controlled Subsequent Activity in the Western United States". International Geology Review. 45 *(7): 575–595. doi:10.2747/0020-6814.45.7.575. ISSN 0020-6814.*
3. Gutscher, Marc-André; Spakman, Wim; Bijwaard, Harmen; Engdahl, E. Robert (2000-10-01). Geodynamics of flat subduction: Seismicity and tomographic constraints from the Andean margin. Tectonics. 19 *(5): 814–833.* doi:10.1029/1999TC001152. ISSN 1944-9194.
4. Rosenbaum, Gideon; Giles, David; Saxon, Mark; Betts, Peter G.; Weinberg, Roberto F.; Duboz, Cecile (2005-10-30). Subduction of the Nazca Ridge and the Inca Plateau: Insights into the formation of ore deposits in Peru. Earth and Planetary Science Letters. 239 *(1–2): 18–32. doi:10.1016/j.epsl.2005.08.003.*
5. Antonijevic, Sanja Knezevic; Wagner, Lara S.; Kumar, Abhash; Beck, Susan L.; Long, Maureen D.; Zandt, George; Tavera, Hernando; Condori, Cristobal (2015-08-13). The role of ridges in the formation and longevity of flat slabs. Nature. 524 *(7564): 212–215.* doi:10.1038/nature14648. ISSN 0028-0836.

6. ^ Jump up to:[a] [b] [c] Manea, Vlad C.; Pérez-Gussinyé, Marta; Manea, Marina (2012-01-01). Chilean flat slab subduction controlled by overriding plate thickness and trench rollback. Geology. 40 *(1): 35–38. doi:10.1130/G32543.1. ISSN 0091-7613.*
7. Uyeda, S.; Sacks, I. Selwyn (1977-01-05). Subduction zones, mid-ocean ridges, oceanic trenches and geodynamicsInterrelationships between volcanism, seismicity, and anelasticity in western South America. Tectonophysics. 37 *(1): 131–139.* doi:10.1016/0040-1951(77)90043-9.
8. Hasegawa, Akira; Sacks, I. Selwyn (1981-06-10). Subduction of the Nazca Plate beneath Peru as determined from seismic observations. Journal of Geophysical Research: Solid Earth. 86 *(B6): 4971–4980.* doi:10.1029/JB086iB06p04971. ISSN 2156-2202.
9. van Hunen, Jeroen; van den Berg, Arie P; Vlaar, Nico J (2004-08-16). Various mechanisms to induce present-day shallow flat subduction and implications for the younger Earth: a numerical parameter study. Physics of the Earth and Planetary Interiors. Plumes and Superplumes. 146 *(1–2): 179–194. doi:10.1016/j.pepi.2003.07.027.*
10. Skinner, Steven M.; Clayton, Robert W. (2013-06-01). The lack of correlation between flat slabs and bathymetric impactors in South America. Earth and Planetary Science Letters. 371–372: 1–5. doi:10.1016/j.epsl.2013.04.013.
11. Rosenbaum, Gideon; Mo, Won (2011-04-01). Tectonic and magmatic responses to the subduction of high bathymetric relief. Gondwana Research. Island Arcs: Their role in growth of accretionary orogens and mineral endowment. 19 *(3): 571–582. doi:10.1016/j.gr.2010.10.007.*
12. Stevenson, D. J.; Turner, J. S. (1977-11-24). Angle of subduction. Nature. 270 *(5635): 334–336.* doi:10.1038/270334a0.
13. Schellart, W. P.; Freeman, J.; Stegman, D. R.; Moresi, L.; May, D. (2007-03-15). Evolution and diversity of subduction zones controlled by slab width. Nature. 446 *(7133): 308–311.* doi:10.1038/nature05615. ISSN 0028-0836.
14. Pennington, Wayne D. (1984-02-20). Geodynamics of Back-Arc Regions The effect of oceanic crustal structure on phase changes and subduction. Tectonophysics. 102 *(1): 377–398.* doi:10.1016/0040-1951(84)90023-4.
15. Alvarado, Patricia; Pardo, Mario; Gilbert, Hersh; Miranda, Silvia; Anderson, Megan; Saez, Mauro; Beck, Susan (2009-06-01). Flat-slab subduction and crustal models for the seismically active Sierras Pampeanas region of Argentina. Geological Society of America Memoirs. 204: 261–278. doi:10.1130/2009.1204(12). ISSN 0072-1069.
16. Gutscher, Marc-André; Maury, René; Eissen, Jean-Philippe; Bourdon, Erwan (2000-06-01). Can slab melting be caused by flat subduction?. *Geology.* 28 *(6): 535–538.* doi:10.1130/0091-7613(2000)282.0.CO;2. ISSN 0091-7613.
17. Humphreys, Eugene (2009-06-01). Relation of flat subduction to magmatism and deformation in the western United States. Geological Society of America Memoirs. 204: 85–98. doi:10.1130/2009.1204(04). ISSN 0072-1069.
18. Gutscher, M. -A; Malavieille, J; Lallemand, S; Collot, J. -Y (1999-05-15).

Tectonic segmentation of the North Andean margin: impact of the Carnegie Ridge collision. Earth and Planetary Science Letters. 168 *(3–4): 255–270.* doi:10.1016/S0012-821X(99)00060-6.

19. Litvak, Vanesa D.; Poma, Stella; Kay, Suzanne Mahlburg (2007-09-01). Paleogene and Neogene magmatism in the Valle del Cura region: New perspective on the evolution of the Pampean flat slab, San Juan province, Argentina. Journal of South American Earth Sciences. 24 *(2–4): 117–137. doi:10.1016/j.jsames.2007.04.002.*
20. Sign In. doi:10.2475/ajs.286.10.737. Retrieved 2016-05-04.
21. Cahill, Thomas; Isacks, Bryan L. (1986-04-01). An apparent double-planed Benioff Zone beneath northern Chile resulting from misidentification of reflected phases. Geophysical Research Letters. 13 *(4): 333–336.* doi:10.1029/GL013i004p00333. ISSN 1944-8007.
22. Espurt, N.; Baby, P.; Brusset, S.; Roddaz, M.; Hermoza, W.; Regard, V.; Antoine, P.-O.; Salas-Gismondi, R.; Bolaños, R. (2007-06-01). How does the Nazca Ridge subduction influence the modern Amazonian foreland basin?. *Geology.* 35 *(6): 515–518. doi:10.1130/G23237A.1. ISSN 0091-7613.*
23. Baby, P.; Guyot, J. L.; Deniaud, Y.; Zubieta, D.; Christophoul, F.; Rivadeneira, M.; Jara, F. (1999-01-01). The High Amazonian Basin: tectonic control and mass balance (PDF). International Symposium MANAUS 99, Hydrological and Geochemical Process in Large-scale Rivers Basins: Manaus (Brazil).
24. Kronberg, B.I.; Fralick, P.W.; Benchimol, R.E. (1998-09-01). Late Quaternary sedimentation and palaeohydrology in the Acre foreland basin, SW Amazonia. Basin Research. 10 *(3). ISSN 1365-2117.*
25. Gutscher, Marc-André (2002-04-01). Andean subduction styles and their effect on thermal structure and interplate coupling. Journal of South American Earth Sciences. Flat-Slab Subduction in the Andes. 15 *(1): 3–10.* doi:10.1016/S0895-9811(02)00002-0.
26. Spencer, Jon E. (1994-01-01). A numerical assessment of slab strength during high- and low-angle subduction and implications for Laramide orogenesis. Journal of Geophysical Research: Solid Earth. 99 *(B5): 9227–9236.* doi:10.1029/94jb00503.
27. Ramos, Victor A.; Folguera, Andrés (2009-01-01). Andean flat-slab subduction through time. Geological Society, London, Special Publications. 327 *(1): 31–54. doi:10.1144/SP327.3. ISSN 0305-8719.*
28. Barazangi, Muawia; Isacks, Bryan L. (1976-11-01). Spatial distribution of earthquakes and subduction of the Nazca plate beneath South America. Geology. 4 *(11): 686–692.* doi:10.1130/0091-7613(1976)42.0.CO;2. ISSN 0091-7613.
29. Dorbath, L.; Dorbath, C.; Jimenez, E.; Rivera, L. (1991-01-01). Seismicity and tectonic deformation in the Eastern Cordillera and the sub-Andean zone of central Peru. Journal of South American Earth Sciences. 4 *(1): 13–24.* doi:10.1016/0895-9811(91)90015-D.
30. Gutscher, M. -A.; Olivet, J. -L.; Aslanian, D.; Eissen, J. -P.; Maury, R. (1999-09-15). The lost inca plateau: cause of flat subduction beneath peru?.

Earth and Planetary Science Letters. 171 *(3): 335–341*. doi:10.1016/S0012-821X(99)00153-3.

31. Pennington, Wayne D. (1981-11-10). Subduction of the Eastern Panama Basin and seismotectonics of northwestern South America. Journal of Geophysical Research: Solid Earth. 86 *(B11): 10753–10770*. doi:10.1029/JB086iB11p10753. ISSN 2156-2202.
32. Muntean, John L.; Cline, Jean S.; Simon, Adam C.; Longo, Anthony A. (2011-02-01). Magmatic-hydrothermal origin of Nevada/'s Carlin-type gold deposits. Nature Geoscience. 4 *(2): 122–127*. doi:10.1038/ngeo1064. ISSN 1752-0894.
33. Abbott, Dallas; Drury, Rebecca; Smith, Walter H. F. (1994-10-01). Flat to steep transition in subduction style. Geology. 22 *(10): 937–940*. doi:10.1130/0091-7613(1994)0222.3.CO;2. ISSN 0091-7613.

CHAPTER 5

Process of Mantle Convection

Chapter Contents

- **Introduction**
- **Continental Drift**
- **Seafloor Spreading**
- **Vine–Matthews–Morley Hypothesis**
- **History**
- **Geomagnetism**

Mantle convection is the process by which heat from the Earth's core is transferred upwards to the surface. It is thought that heating of the mantle by the core creates convection cells in which hot mantle material rises, cooling as it goes, toward the crust until it reaches less dense material, at which point it spreads out then descends. Similar processes can be observed in any fluid above a hot or warm surface — for example, the atmosphere. Mantle convection is assumed to be responsible for plate tectonics and continental drift as well as volcanism.

INTRODUCTION

Mantle convection is the slow creeping motion of Earth's solid silicate mantle caused by convection currents carrying heat from the interior of the Earth to the surface.The Earth's surface lithosphere, which rides atop the asthenosphere (the two components of the upper mantle), is divided into a number of plates that are continuously being created and consumed at their opposite plate boundaries. Accretion occurs as mantle is added to the growing edges of a plate, associated with seafloor spreading. This hot added material cools down by conduction and convection of heat. At the consumption edges of the plate, the material has thermally contracted to become dense, and it sinks under its own weight in the process of subduction usually at an ocean trench. This *subducted* material sinks through the Earth's interior. Some subducted material appears to reach the lower mantle, while in other regions, this material is impeded from sinking further, possibly due to a phase transition from spinel to silicate perovskite and magnesiowustite, an endothermic reaction.

The subducted oceanic crust triggers volcanism, although the basic mechanisms are varied. Volcanism may occur due to processes that add buoyancy to partially melted mantle causing an upward flow due to a decrease in density of the partial melt.

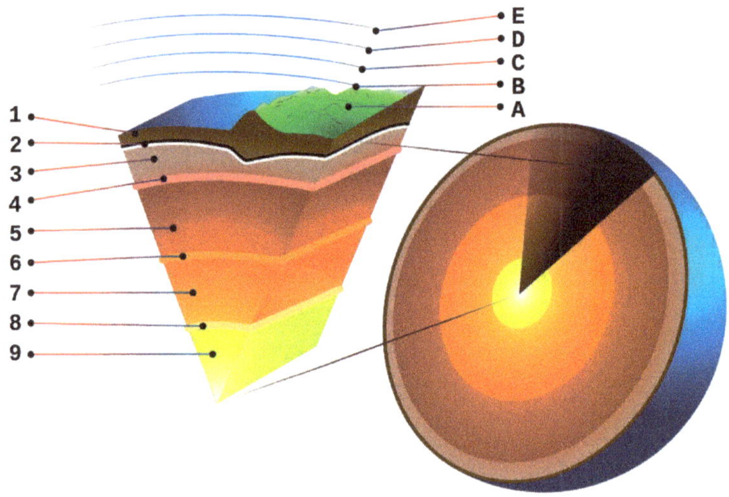

Earth cross-section showing location of upper (3) and lower (5) mantle.

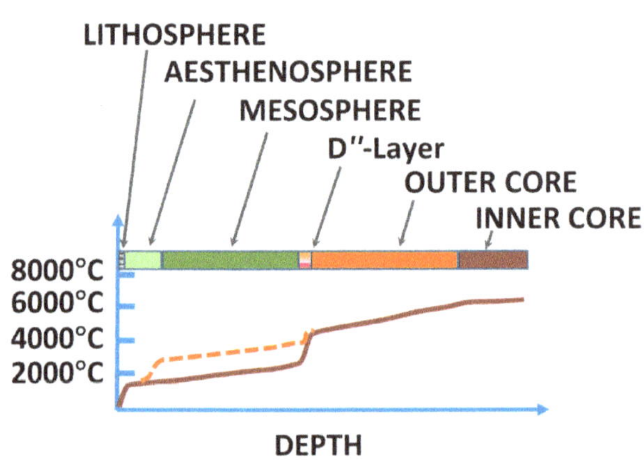

Calculated Earth's temperature vs. depth. Dashed curve: Layered mantle convection; Solid curve: Whole mantle convection.

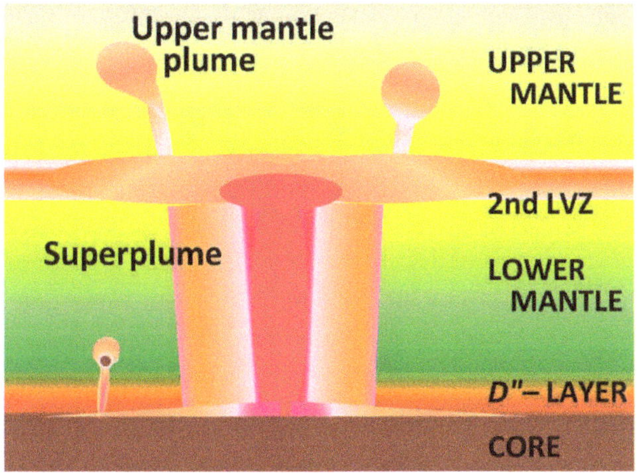

A superplume generated by cooling processes in the mantle.

Secondary forms of convection that may result in surface volcanism are postulated to occur as a consequence of intraplate extension and mantle plumes.

It is because the mantle can convect that the tectonic plates are able to move around the Earth's surface.

Mantle convection seems to have been much more active during the Hadean period, resulting in gravitational sorting of heavier molten iron, and nickel elements and sulphides in the core, and lighter silicate minerals in the mantle.

Types of Convection

During the late 20th century, there was significant debate within the geophysics community as to whether convection is likely to be 'layered' or 'whole'. Although elements of this debate still continue, results from seismic tomography, numerical simulations of mantle convection and examination of Earth's gravitational field are all beginning to suggest the existence of 'whole' mantle convection, at least at the present time. In this model, cold, subducting oceanic lithosphere descends all the way from the surface to the core-mantle boundary (CMB) and hot plumes rise from the CMB all the way to the surface. This picture is strongly based on the results of global seismic tomography models, which typically show slab and plume-like anomalies crossing the mantle transition zone.

Although it is now well accepted that subducting slabs cross the mantle transition zone and descend into the lower mantle, debate about the existence and continuity of plumes persists, with important implications for the style of mantle convection. This debate is linked to the controversy regarding whether intraplate volcanism is caused by shallow, upper-mantle processes or by plumes from the lower mantle. Many geochemistry studies have argued that the lavas erupted in intraplate areas are different in composition from shallow-derived mid ocean ridge basalts (MORB). Specifically, they typically have elevated Helium-3 - Helium-4 ratios. Being a primordial nuclide, Helium-3 is not naturally produced on earth. It also quickly escapes from earth's atmosphere when erupted. The elevated He-3/He-4 ratio of Ocean Island Basalts (OIBs) suggest that they must be sources from a part of the earth that has not previously been melted and reprocessed in the same way as MORB source has been. This has been interpreted as their originating from a different, less well-mixed, region, suggested to be the lower mantle. Others, however, have pointed out that geochemical differences could indicate the inclusion of a small component of near-surface material from the lithosphere.

Speed of Convection

Typical mantle convection speed is 20 mm/yr near the crust but can vary quite a bit. The small scale convection in the upper mantle is much faster than the convection near the core. (See whole-mantle discussion above.) A single shallow convection cycle takes on the order of 50 million years, though deeper convection can be closer to 200 million years.

Creep in the Mantle

Since the mantle is primarily composed of olivine ((Mg,Fe)2SiO4), the rheological characteristics of the mantle are largely those of olivine. Additionally, due to the varying temperatures and pressures between the lower and upper mantle, a variety of creep processes can occur with dislocation creep dominating in the lower mantle and diffusional creep occasionally dominating in the upper mantle. However, there is a large transition region in creep processes between the upper and lower mantle and even within each section, creep properties can change strongly with location and thus temperature and pressure. In the power law creep regions, the creep equation fitted to data with n = 3-4 is standard.

The strength of olivine not only scales with its melting temperature, but also is very sensitive to water and silica content. The solidus depression by impurities, primarily Ca, Al, and Na, and pressure affects creep behavior and thus contributes to the change in creep mechanisms with location. While creep behavior is generally plotted as homologous temperature versus stress, in the case of the mantle it is often more useful to look at the pressure dependence of stress. Though stress is simple force over area, defining the area is difficult in geology. Equation 1 demonstrates the pressure dependence of stress. Since it is very difficult to simulate the high pressures in the mantle (1MPa at 300–400 km), the low pressure laboratory data is usually extrapolated to high pressures by applying creep concepts from metallurgy.

$$(1)\quad \left(\frac{\partial \ln \sigma}{\partial P}\right)_{T,\dot{\epsilon}} = \left(\frac{1}{TT_m}\right) * \left(\frac{\partial \ln \sigma}{\partial(1/T)}\right)_{P,\dot{\epsilon}} * \frac{dT_m}{dP}$$

Most of the mantle has homologous temperatures of 0.65-0.75 and experiences strain rates of 10^-14 – 10^-16 1/s. Stresses in mantle are dependent on density, gravity, thermal expansion coefficients, temperature differences driving convection, and distance convection occurs over, all of which give stresses around a fraction of 3-30MPa. Due to the large grain sizes (at low stresses as high as several mm), it is unlikely that Nabarro-Herring (NH) creep truly dominates. Given the large grain sizes, dislocation creep tends to dominate. 14 MPa is the stress below which diffusional creep dominates and above which power law creep dominates at 0.5Tm of olivine. Thus, even for relatively low temperatures, the stress diffusional creep would operate at is too low for realistic conditions. Though the power law creep rate increases with increasing water content due to weakening, reducing activation energy of diffusion and thus increasing the NH creep rate, NH is generally still not large enough to dominate. Nevertheless, diffusional creep can dominate in very cold or deep parts of the upper mantle. Additional deformation in the mantle can be attributed to transformation enhanced ductility. Below 400 km, the olivine undergoes a pressure induced phase transformation into spinel and can cause more deformation due to the increased ductility. Further evidence for the dominance of power law creep comes from preferred lattice orientations as a result of deformation. Under dislocation creep, crystal structures reorient into lower stress orientations. This does not happen under diffusional creep, thus observation of preferred orientations in samples lends credence to the dominance of dislocation creep.

CONTINENTAL DRIFT

The continental drift of the last 150 million years

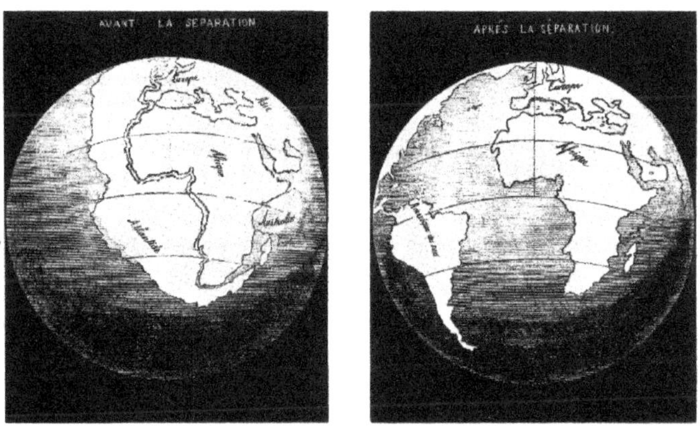

Antonio Snider-Pellegrini's Illustration of the closed and opened Atlantic Ocean (1858).

Continental drift is the movement of the Earth's continents relative to each other, thus appearing to "drift" across the ocean bed. The speculation that continents might have 'drifted' was first put forward by Abraham Ortelius in 1596. The concept was independently and more fully developed by Alfred Wegener in 1912, but his theory was rejected by some for lack of a mechanism (though this was supplied later by Arthur Holmes) and others because of prior theoretical commitments. The idea of continental drift has been subsumed by the theory of plate tectonics, which explains how the continents move.

In 1858 Antonio Snider-Pellegrini created two maps demonstrating how the American and African continents might have once fit together.

Early History

Abraham Ortelius (Ortelius 1596), Theodor Christoph Lilienthal (1756), Alexander von

Humboldt (1801 and 1845),Antonio Snider-Pellegrini (Snider-Pellegrini 1858), and others had noted earlier that the shapes of continents on opposite sides of the Atlantic Ocean (most notably, Africa and South America) seem to fit together. W. J. Kious described Ortelius' thoughts in this way:

Abraham Ortelius in his work Thesaurus Geographicus ... suggested that the Americas were torn away from Europe and Africa ... by earthquakes and floods and went on to say: The vestiges of the rupture reveal themselves, if someone brings forward a map of the world and considers carefully the coasts of the three [continents].

Writing in 1889, Alfred Russel Wallace remarks "It was formerly a very general belief, even amongst geologists, that the great features of the earth's surface, no less than the smaller ones, were subject to continual mutations, and that during the course of known geological time the continents and great oceans had again and again changed places with each other." He quotes Charles Lyell as saying "Continents, therefore, although permanent for whole geological epochs, shift their positions entirely in the course of ages" and claims that the first to throw doubt on this was James Dwight Dana in 1849.

In his *Manual of Geology*, 1863, Dana says "The continents and oceans had their general outline or form defined in earliest time. This has been proved with respect to North America from the position and distribution of the first beds of the Silurian - those of the Potsdam epoch. ... and this will probably prove to the case in Primordial time with the other continents also".Dana was enormously influential in America - his *Manual of Mineralogy* is still in print in revised form - and the theory became known as *Permanence theory*.

This appeared to be confirmed by the exploration of the deep sea beds conducted by the Challenger expedition, 1872-6, which showed that contrary to expectation, land debris brought down by rivers to the ocean is deposited comparatively close to the shore in what is now known as the continental shelf. This suggested that the oceans were a permanent feature of the earth's surface, and did not change places with the continents.

Wegener and his Predecessors

The speculation that the American continents had once formed a single landmass with Europe and Asia before assuming the present shapes and positions was suggested by several scientists before Alfred Wegener's 1912 paper. Although Wegener's theory was formed independently and was more complete than those of his predecessors, Wegener later credited a number of past authors with similar ideas: Franklin Coxworthy (between 1848 and 1890), Roberto Mantovani (between 1889 and 1909), William Henry Pickering (1907) and Frank Bursley Taylor (1908).In addition, Eduard Suess had proposed a supercontinent Gondwana in 1885 and the Tethys Ocean in 1893, assuming a land-bridge between the present continents submerged in the form of a geosyncline, and John Perry had written an 1895 paper proposing that the earth's interior was fluid, and disagreeing with Lord Kelvin on the age of the earth.

For example: the similarity of southern continent geological formations had led Roberto Mantovani to conjecture in 1889 and 1909 that all the continents had once been joined into a supercontinent; Wegener noted the similarity of Mantovani›s and his own maps of the former positions of the southern continents. In Mantovani›s conjecture, through volcanic activity due to thermal expansion this continent broke and the new continents drifted away from each other because of further expansion of the rip-zones, where the oceans now lie. This led Mantovani to propose an Expanding Earth theory which has since been shown to be incorrect.

Alfred Wegener.

Continental drift without expansion was proposed by Frank Bursley Taylor, who suggested in 1908 (published in 1910) that the continents were moved into their present positions by a process of continental creep. In a later paper he proposed that this occurred by their being dragged towards the equator by tidal forces during the hypothesized capture of the moon in the Cretaceous, resulting in general crustal creep toward the equator. Although his proposed mechanism was wrong, he was the first to realize the insight that one of the effects of continental motion would be the formation of mountains, and attributed the formation of the Himalayas to the collision between the Indian subcontinent with Asia. Wegener said that of all those theories, Taylor›s, although not fully developed, had the most similarities to his own. In the mid-20th century, the theory of continental drift was referred to as the "Taylor-Wegener hypothesis", although this terminology eventually fell out of common use.

Alfred Wegener first presented his hypothesis to the German Geological Society on January 6, 1912. His hypothesis was that the continents had once formed a single landmass, called Pangea, before breaking apart and drifting to their present locations.

Wegener was the first to use the phrase continental drift (1912, 1915) (in German die Verschiebung der Kontinente – translated into English in 1922) and formally publish the hypothesis that the continents had somehow drifted apart. Although he presented much evidence for continental drift, he was unable to provide a convincing explanation for the physical processes which might have caused this drift. His suggestion that the continents had been pulled apart by the centrifugal pseudoforce (*Polflucht*) of the Earth's rotation or by a small component of astronomical precession was rejected as calculations showed that the force was not sufficient. The Polflucht hypothesis was also studied by Paul Sophus Epstein in 1920 and found to be implausible.

Rejection of Wegener›s theory, 1910s–1950s

The theory of continental drift was not accepted for many years. One problem was that a plausible driving force was missing.A second problem was that Wegener's estimate of the velocity of continental motion, 250 cm/year, was implausibly high.(The currently accepted rate for the separation of the Americas from Europe and Africa is about 2.5 cm/year). And it did not help that Wegener was not a geologist. Other geologists also believed that the evidence that Wegener had provided was not sufficient. It is now accepted that the plates carrying the continents do move across the Earth's surface, although not as fast as Wegener believed; ironically one of the chief outstanding questions is the one Wegener failed to resolve: what is the nature of the forces propelling the plates?

The British geologist Arthur Holmes championed the theory of continental drift at a time when it was deeply unfashionable. He proposed in 1931 that the Earth's mantle contained convection cells that dissipated radioactive heat and moved the crust at the surface. His *Principles of Physical Geology*, ending with a chapter on continental drift, was published in 1944.

David Attenborough, who attended university in the second half of the 1940s, recounted an incident illustrating its lack of acceptance then: "I once asked one of my lecturers why he was not talking to us about continental drift and I was told, sneeringly, that if I could prove there was a force that could move continents, then he might think about it. The idea was moonshine, I was informed."

Geological maps of the time showed huge land bridges spanning the Atlantic and Indian oceans to account for the similarities of fauna and flora and the divisions of the Asian continent in the Permian era but failing to account for glaciation in India, Australia and South Africa.

As late as 1953 – just five years before Carey introduced the theory of plate tectonics – the theory of continental drift was rejected by the physicist Scheidegger on the following grounds.

- First, it had been shown that floating masses on a rotating geoid would collect at the equator, and stay there. This would explain one, but only one, mountain building episode between any pair of continents; it failed to account for earlier orogenic episodes.
- Second, masses floating freely in a fluid substratum, like icebergs in the ocean, should be in isostatic equilibrium (in which the forces of gravity and buoyancy are in balance). But gravitational measurements showed that many areas are not in isostatic equilibrium.

- Third, there was the problem of why some parts of the Earth›s surface (crust) should have solidified while other parts were still fluid. Various attempts to explain this foundered on other difficulties.

Geophysicist Jack Oliver is credited with providing seismologic evidence supporting plate tectonics which encompassed and superseded continental drift with the article "Seismology and the New Global Tectonics", published in 1968, using data collected from seismologic stations, including those he set up in the South Pacific.

It is now known that there are two kinds of crust: continental crust and oceanic crust. Continental crust is inherently lighter and its composition is different from oceanic crust, but both kinds reside above a much deeper plastic mantle. Oceanic crust is created at spreading centers, and this, along with subduction, drives the system of plates in a chaotic manner, resulting in continuous orogeny and areas of isostatic imbalance. The theory of plate tectonics explains all this, including the movement of the continents, better than Wegener's theory.

Evidence of Continental Drift

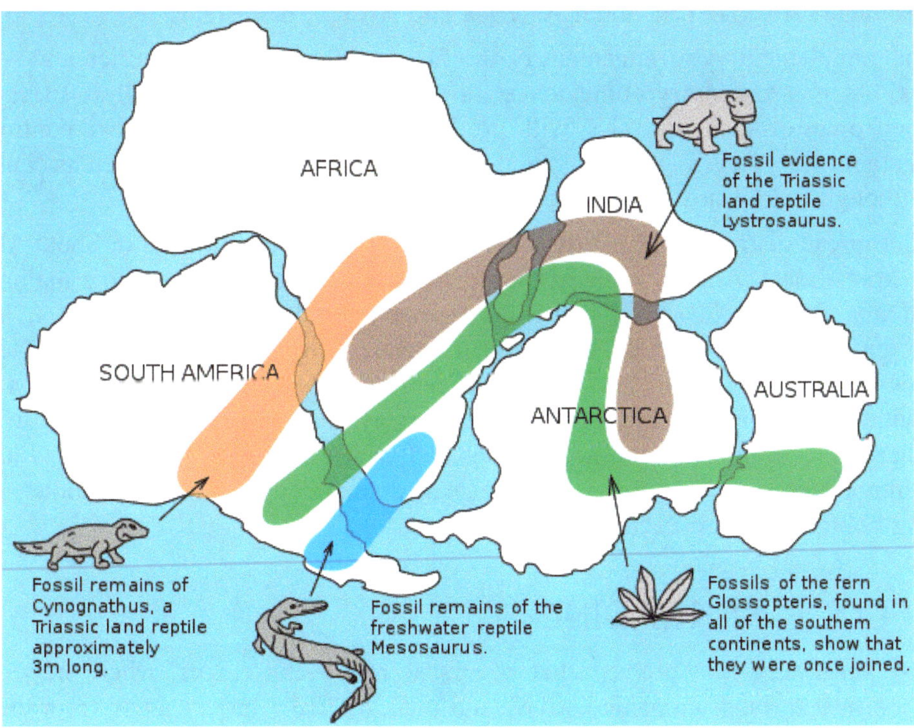

Fossil patterns across continents (Gondwanaland).

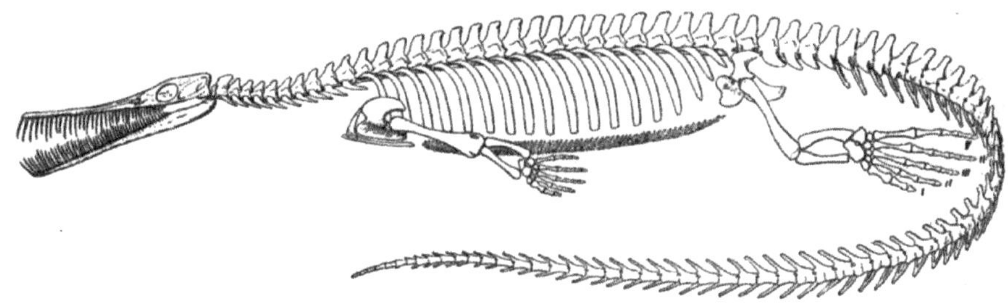

Mesosaurus skeleton, MacGregor, 1908.

Evidence for the movement of continents on tectonic plates is now extensive. Similar plant and animal fossils are found around the shores of different continents, suggesting that they were once joined. The fossils of *Mesosaurus*, a freshwater reptile rather like a small crocodile, found both in Brazil and South Africa, are one example; another is the discovery of fossils of the land reptile *Lystrosaurus* in rocks of the same age at locations in Africa, India, and Antarctica. There is also living evidence—the same animals being found on two continents. Some earthworm families (e.g. Ocnerodrilidae, Acanthodrilidae, Octochaetidae) are found in South America and Africa, for instance.

The complementary arrangement of the facing sides of South America and Africa is obvious, but is a temporary coincidence. In millions of years, slab pull and ridge-push, and other forces of tectonophysics, will further separate and rotate those two continents. It was this temporary feature which inspired Wegener to study what he defined as continental drift, although he did not live to see his hypothesis generally accepted.

Widespread distribution of Permo-Carboniferous glacial sediments in South America, Africa, Madagascar, Arabia, India, Antarctica and Australia was one of the major pieces of evidence for the theory of continental drift. The continuity of glaciers, inferred from oriented glacial striations and deposits called tillites, suggested the existence of the supercontinent of Gondwana, which became a central element of the concept of continental drift. Striations indicated glacial flow away from the equator and toward the poles, based on continents› current positions and orientations, and supported the idea that the southern continents had previously been in dramatically different locations, as well as being contiguous with each other.

SEAFLOOR SPREADING

Seafloor spreading is a process that occurs at mid-ocean ridges, where new oceanic crust is formed through volcanic activity and then gradually moves away from the ridge. Seafloor spreading helps explain continental drift in the theory of plate tectonics. When oceanic plates diverge, tensional stress causes fractures to occur in the lithosphere. Basaltic magma rises up the fractures and cools on the ocean floor to form new sea floor. Older rocks will be found farther away from the spreading zone while younger rocks will be found nearer to the spreading zone.

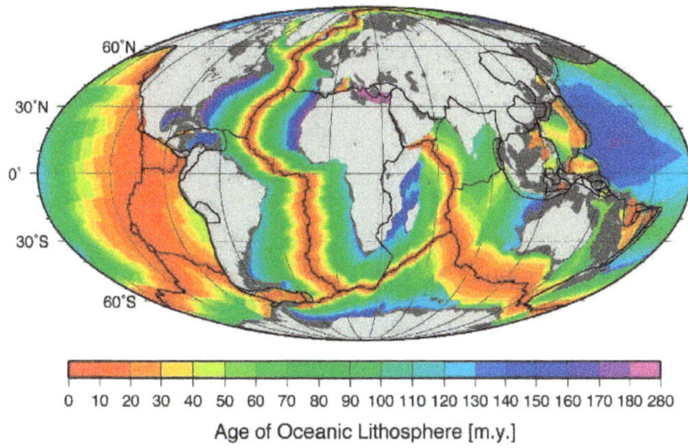

Age of oceanic lithosphere; youngest (red) is along spreading centers.

Earlier theories (e.g. by Alfred Wegener and Alexander du Toit) of continental drift were that continents "ploughed" through the sea. The idea that the seafloor itself moves (and also carries the continents with it) as it expands from a central axis was proposed by Harry Hess from Princeton University in the 1960s. The theory is well accepted now, and the phenomenon is known to be caused by convection currents in the asthenosphere, which is ductile, or plastic, and the brittle lithosphere.

Incipient Spreading

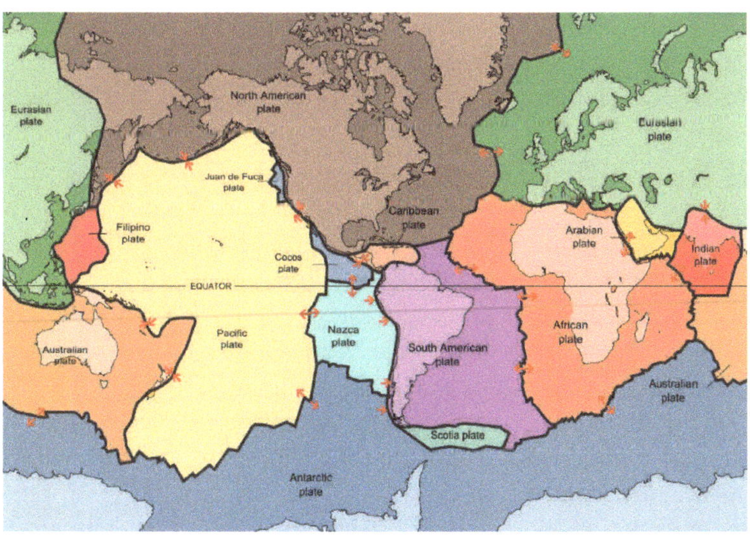

Plates in the crust of the earth, according to the plate tectonics theory.

In the general case, sea floor spreading starts as a rift in a continental land mass, similar to the Red Sea-East Africa Rift System today. The process starts with heating at the base of the continental crust which causes it to become more plastic and less dense. Because less dense objects rise in relation to denser objects, the area being heated becomes a broad dome (see isostasy). As the crust bows upward, fractures occur that gradually grow into rifts. The typical rift system consists of three rift arms at approximately 120 degree angles. These areas are named triple junctions and can be found in several places across the world today. The separated margins of the continents evolve to form passive margins. Hess› theory was that new seafloor is formed when magma is forced upward toward the surface at a mid-ocean ridge.

If spreading continues past the incipient stage described above, two of the rift arms will open while the third arm stops opening and becomes a ‹failed rift›. As the two active rifts continue to open, eventually the continental crust is attenuated as far as it will stretch. At this point basaltic oceanic crust begins to form between the separating continental fragments. When one of the rifts opens into the existing ocean, the rift system is flooded with seawater and becomes a new sea. The Red Sea is an example of a new arm of the sea. The East African rift was thought to be a "failed" arm that was opening somewhat more slowly than the other two arms, but in 2005 the Ethiopian Afar Geophysical Lithospheric Experiment reported that in the Afar region last September, a 60 km fissure opened as wide as eight meters. During this period of initial flooding the new sea is sensitive to changes in climate and eustasy. As a result, the new sea will evaporate (partially or completely) several times before the elevation of the rift valley has been lowered to the point that the sea becomes stable. During this period of evaporation large evaporite deposits will be made in the rift valley. Later these deposits have the potential to become hydrocarbon seals and are of particular interest to petroleum geologists.

Sea floor spreading can stop during the process, but if it continues to the point that the continent is completely severed, then a new ocean basin is created. The Red Sea has not yet completely split Arabia from Africa, but a similar feature can be found on the other side of Africa that has broken completely free. South America once fit into the area of the Niger Delta. The Niger River has formed in the failed rift arm of the triple junction.

Continued Spreading and Subduction

As new seafloor forms and spreads apart from the mid-ocean ridge it slowly cools over time. Older seafloor is therefore colder than new seafloor, and older oceanic basins deeper than new oceanic basins due to isostasy. If the diameter of the earth remains relatively constant despite the production of new crust, a mechanism must exist by which crust is also destroyed. The destruction of oceanic crust occurs at subduction zones where oceanic crust is forced under either continental crust or oceanic crust. Today, the Atlantic basin is actively spreading at the Mid-Atlantic Ridge. Only a small portion of the oceanic crust produced in the Atlantic is subducted. However, the plates making up the Pacific Ocean are experiencing subduction along many of their boundaries which causes the volcanic activity in what has been termed the Ring of Fire of the Pacific Ocean. The Pacific is also home to one of the world's most active spreading centres (the East Pacific Rise) with spreading rates of up to 13 cm/yr. The Mid-Atlantic Ridge is a textbook slow-spreading centre, while the East Pacific Rise is used as an example of fast spreading. The differences in spreading rates affect not only the geometries of the ridges but also the geochemistry of the basalts that are produced. Since the new oceanic basins are shallower than the old oceanic basins, the total capacity of the world›s ocean basins decreases during times of active sea floor spreading. During the opening of the Atlantic Ocean, sea level was so high that a Western Interior Seaway formed across North America from the Gulf of Mexico to the Arctic Ocean.

Debate and Search for Mechanism

At the Mid-Atlantic Ridge (and in other areas), material from the upper mantle rises through the faults between oceanic plates to form new crust as the plates move away from each other, a phenomenon first observed as continental drift. When Alfred Wegener first presented a hypothesis of continental drift in 1912, he suggested that continents ploughed through the ocean crust. This was impossible: oceanic crust is both more dense and more rigid than continental crust. Accordingly, Wegener›s theory wasn›t taken very seriously, especially in the United States.

Since then, it has been shown that the motion of the continents is linked to seafloor spreading. In the 1960s, the past record of geomagnetic reversals was noticed by observing the magnetic stripe "anomalies" on the ocean floor. This results in broadly evident "stripes" from which the past magnetic field polarity can be inferred by looking at the data gathered from simply towing a magnetometer on the sea surface or from an aircraft. The stripes on one side of the mid-ocean ridge were the mirror image of those on the other side. The seafloor must have originated on the Earth's great fiery welts, like the Mid-Atlantic Ridge and the East Pacific Rise.

The driver for seafloor spreading in plates with active margins is the weight of the cool, dense, subducting slabs that pull them along. The magmatism at the ridge is considered to be "passive upswelling", which is caused by the plates being pulled apart under the weight of their own slabs. This can be thought of as analogous to a rug on a table with little friction: when part of the rug is off of the table, its weight pulls the rest of the rug down with it.

Sea Floor Global Topography: half-space Model

To first approximation, sea floor global topography in areas without significant subduction can be estimated by the half-space model. In this model, the seabed height is determined by the oceanic lithosphere temperature, due to thermal expansion. Oceanic lithosphere is continuously formed at a constant rate at the mid-ocean ridges. The source of the lithosphere has a half-plane shape ($x = 0$, $z < 0$) and a constant temperature T_1. Due to its continuous creation, the lithosphere at $x > 0$ is moving away from the ridge at a constant velocity v, which is assumed large compared to other typical scales in the problem. The temperature at the upper boundary of the lithosphere ($z=0$) is a constant $T_0 = 0$. Thus at $x = 0$ the temperature is the Heaviside step function $T_1 \cdot \Theta(-z)$. Finally, we assume the system is at a quasi-steady state, so that the temperature distribution is constant in time, i.e. $T=T(x,z)$. By calculating in the frame of reference of the moving lithosphere (velocity v), which have spatial coordinate $x' = x-vt$, we may write $T = T(x',z,t)$ and use the heat equation: $\frac{\partial T}{\partial t} = \kappa \nabla^2 T = \kappa \frac{\partial^2 T}{\partial^2 z} + \kappa \frac{\partial^2 T}{\partial^2 x'}$ where κ is the thermal diffusivity of the mantle lithosphere.

Since T depends on x› and t only through the combination $x = x' + vt$, we have: $\frac{\partial T}{\partial x'} = \frac{1}{v} \cdot \frac{\partial T}{\partial t}$

Thus: $\frac{\partial T}{\partial t} = \kappa \nabla^2 T = \kappa \frac{\partial^2 T}{\partial^2 z} + \frac{\kappa}{v^2} \frac{\partial^2 T}{\partial^2 t}$.

We now use the assumption that v is large compared to other scales in the problem; we therefore neglect the last term in the equation, and get a 1-dimensional diffusion equation: $\frac{\partial T}{\partial t} = \kappa \frac{\partial^2 T}{\partial^2 z}$ with the initial conditions $T(t = 0) = T_1 \cdot \Theta(-z)$.

The solution for $z \leq 0$ is given by the error function $\mathbf{erf}$:

$$T(x', z, t) = T_1 \cdot \mathbf{erf}\left(\frac{z}{2\sqrt{\kappa t}}\right)$$

Due to the large velocity, the temperature dependence on the horizontal direction is negligible, and the height at time t (i.e. of sea floor of age t) can be calculated by integrating the thermal expansion over z:

$$h(t) = h_0 + \alpha_{\text{eff}} \int_0^\infty [T(z) - T_1] dz = h_0 - \frac{2}{\sqrt{\pi}} \alpha_{\text{eff}} T_1 \sqrt{\kappa t}$$

where α_{eff} is the effective volumetric thermal expansion coefficient, and h_0 is the mid-ocean ridge height (compared to some reference).

Note that the assumption the v is relatively large is equivalently to the assumption that the thermal diffusivity K is small compared to L^2/T, where L is the acean width (from mid-ocean ridges to continental shelf) and T is its age.

The effective thermal expansion coefficient α_{eff} is different from the usual thermal expansion coefficient α due to isostasic effect of the change in water column height above the lithosphere as it expands or retracts. Both coefficients are related by:

$$\alpha_{\text{eff}} = \alpha \cdot \frac{\rho}{\rho - \rho_w}$$

where $\rho \sim 3.3 g/cm^3$ is the rock density and $\rho_0 = 1 g/cm^3$ is the density of water.

By substituting the parameters by their rough estimates: $\kappa \sim 8 \cdot 10^{-7}$ m^2/s, $\alpha \sim 4 \cdot 10^{-5}$ °C^{-1} and T_1 ~1220 °C (for the Atlantic and Indian oceans) or ~1120 °C (for the eastern Pacific), we have:

$$h(t) \sim h_0 - 350\sqrt{t}$$

for the eastern Pacific Ocean, and:

$$h(t) \sim h_0 - 390\sqrt{t}$$

for the Atlantic and Indian Ocean, where the height is in meters and time is in millions of years. To get the dependence on x, one must substitute t = x/v ~ Tx/L, where L is the distance between the ridge to the continental shelf (roughly half the ocean width), and T is the ocean age.

VINE–MATTHEWS–MORLEY HYPOTHESIS

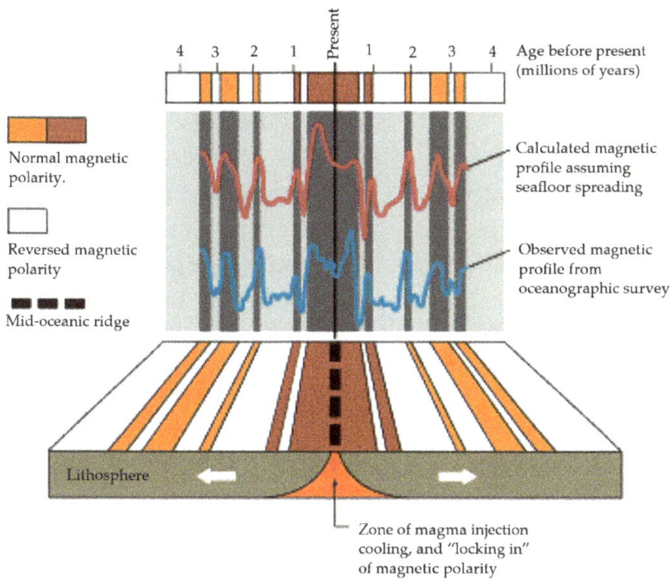

The observed magnetic profile for the sea floor around a mid-oceanic ridge agrees closely with the profile predicted by the Vine–Matthews–Morley hypothesis.

The **Vine–Matthews–Morley hypothesis**, also known as the **Morley–Vine–Matthews hypothesis**, was the first key scientific test of the seafloor spreading theory of continental drift and plate tectonics.

HISTORY

Harry Hess proposed the sea-floor spreading hypothesis in 1960 (published in 1962). According to Hess, seafloor was created at mid-oceanic ridges by the convection of the earth's mantle, pushing and spreading the older crust away from the ridge. Geophysicist Frederick John Vine and the Canadian geologist Lawrence W. Morley independently realized that if Hess's seafloor spreading theory was correct, then the rocks surrounding the mid-oceanic ridges should show symmetric patterns of magnetization reversals using newly collected magnetic surveys. Both of Morley's letters to *Nature* (February 1963) and *Journal of Geophysical Research* (April 1963) were rejected, hence Vine and his PhD adviser at Cambridge University, Drummond Hoyle Matthews, were first to publish the theory in 1963. Some colleagues were skeptical of the hypothesis because of the numerous assumptions made—seafloor spreading, geomagnetic reversals, and remnant magnetism—all hypotheses that were still not widely accepted.The Vine-Matthews hypothesis describes the magnetic reversals of oceanic crust. Further proof of this hypothesis came from Cox et al. (1967) when he measured the remnant magnetization of lavas from land sites. Walter C. Pitman offered further evidence with a remarkably symmetric profile from the Pacific-Antarctic Ridge.

GEOMAGNETISM

The Vine-Matthews hypothesis correlates the symmetric magnetic patterns seen on the seafloor with geomagnetic field reversals. At mid-ocean ridges, new crust is created by the injection, extrusion, and solidification of magma. After the magma has cooled through the Curie point, ferromagnetism becomes possible and the magnetic minerals in the newly formed crust orient themselves with the current background geomagnetic field. Lithospheric creation at the ridge is considered continuous and symmetrical as the new crust pushes the old crust laterally and equally on either side of the ridge. Therefore, as geomagnetic reversal occur, the crust on either side of the ridge will contain a record of remnant magnetizations of normal or reversed magnetizations in comparison to the current geomagnetic field. The ridge crest is analogous to "twin-headed tape recorder", recording the Earth's magnetic history. The intensity of the remnant magnetization is greater than the induced magnetization. The shape of the magnetic anomaly is controlled by the combination of the orientation of its total magnetization, the summation of the induced magnetization and remnant magnetization. Consequently, the shape of the magnetic anomaly is controlled predominately by the primary remnant vector. Blocks of crust at high latitudes have magnetic vectors that dip steeply downward in a normal magnetic field. However, close to the magnetic south pole, the magnetic vectors dip steeply downward in a reversed magnetic field and inclined steeply upwards in a normal magnetic field. There are positive magnetic anomalies over normally magnetized blocks and negative anomalies over reversed blocks. Local anomalies with a short period also exist, but are considered to be correlated with bathymetry. In low latitudes, the vectors are shallowly dipping and dipolar. This confuses the positive and negative anomalies. At the magnetic equator, the magnetic field is horizontal; negative anomalies are related to normally magnetized blocks and positive anomalies with reversed. The amplitude of the anomaly, magnetic field strength, and the magnitude of the remanence all decrease from the poles to the equator. The orientation of the ridge affects the anomaly shape and amplitude. The component of the vector that effects the anomaly is at a maximum when the ridge is east-west and the magnetic profile is north-south. The anomalies are most apparent at high magnetic latitudes, north-south trending ridges at all latitudes, and east-west trending ridges at the magnetic equator.

REFERENCES

1. Antonio Snider-Pellegrini, La Création et ses mystères dévoilés (Creation and its mysteries revealed) (Paris, France: Frank et Dentu, 1858), plates 9 and 10 (between pages 314 and 315).
2. USGS: Historical perspective on plate tectonics, last updated 19 September 2011
3. Oreskes 2002, p. 324.

4. Romm, James (February 3, 1994), A New Forerunner for Continental Drift, Nature, **367** *(6462): 407–408, Bibcode:*1994Natur.367..407R, doi:10.1038/367407a0.
5. Schmeling, Harro (2004). Geodynamik (PDF) (in German). University of Frankfurt.
6. Brusatte, Stephen, Continents Adrift and Sea-Floors Spreading: The Revolution of Plate Tectonics (PDF)
7. Kious, W. J.; Tilling, R. I. (February 2001) [1996], Historical perspective, This Dynamic Earth: the Story of Plate Tectonics (Online ed.), U.S. Geological Survey, ISBN 0-16-048220-8, retrieved 2008-01-29
8. Wallace, Alfred Russel (1889), 12, Darwinism ..., Macmillan, p. 341
9. Lyell, Charles (1872), Principles of Geology ... (11 ed.), John Murray, p. 258
10. Dana, James D. (1863), Manual of Geology, Theodore Bliss & Co, Philadelphia, p. 732
11. Oreskes, Naomi (2002), Continental Drift (PDF), archived from the original on Feb 4, 2012
12. Wegener, Alfred (6 January 1912), Die Herausbildung der Grossformen der Erdrinde (Kontinente und Ozeane), auf geophysikalischer Grundlage (PDF), Petermanns Geographische Mitteilungen, **63**: 185–195, 253–256, 305–309.
13. Wegener, A. (1966) [1929], The Origin of Continents and Oceans, Courier Dover Publications, ISBN 0-486-61708-4
14. Wegener, A. (1929), Die Entstehung der Kontinente und Ozeane (4 ed.), Braunschweig: Friedrich Vieweg & Sohn Akt. Ges.
15. Coxworthy, Franklin (1924). Electrical Condition; Or, How and where Our Earth was Created. J.S. Phillips. Retrieved December 6, 2014.
16. Pickering, W.H (1907), The Place of Origin of the Moon - The Volcani Problems, Popular Astronomy: 274–287, Bibcode:1907PA.....15..274P
17. Frank Bursley Taylor (June 3, 1910) "Bearing of the Tertiary mountain belt on the origin of the earth's plan", Bulletin of the Geological Society of America, **21** : 179–226.
18. Eduard Suess, Das Antlitz der Erde (The Face of the Earth), vol. 1 (Leipzig, (Germany): G. Freytag, 1885), page 768.From p. 768: Wir nennen es Gondwána-Land, nach der gemeinsamen alten Gondwána-Flora, ... (We name it Gondwána-Land, after the common ancient flora of Gondwána ...)
19. Edward Suess (March 1893) "Are ocean depths permanent?, Natural Science: A Monthly Review of Scientific Progress (London), **2** : 180- 187. From page 183: "This ocean we designate by the name "Tethys", after the sister and consort of Oceanus. The latest successor of the Tethyan Sea is the present Mediterranean."
20. Perry, John (1895) "On the age of the earth", Nature, **51** : 224-227, 341-342, 582-585.
21. Mantovani, R. (1889), Les fractures de l›écorce terrestre et la théorie de Laplace, Bull. Soc. Sc. Et Arts Réunion: 41–53

22. Mantovani, R. (1909), L›Antarctide, Je m'instruis. La science pour tous, **38**: 595–597
23. Scalera, G. (2003), Roberto Mantovani an Italian defender of the continental drift and planetary expansion, in Scalera, G.; Jacob, K.-H., Why expanding Earth? – A book in honour of O.C. Hilgenberg, Rome: Istituto Nazionale di Geofisica e Vulcanologia, pp. 71–74
24. Obituary: Frank Bursley Taylor (1860 – 1938), Proceedings of the American Academy, Vol. 75, No. 6 (Dec. 1944), American Academy of Arts and Sciences, pp. 176-178 (retrieved 20 Oct. 2015)
25. Taylor, F.B. (1910), Bearing of the tertiary mountain belt on the origin of the earth›s plan, GSA Bulletin, **21** *(2): 179–226, doi:10.1130/1052-5173(2005)015[29b: WTCCA]2.0.CO;2*
26. Henry R. Frankel, Wegener and Taylor develop their theories of continental drift, in The Continental Drift Controversy: Wegener and the Early Debate Volume 1, Wegener and the Early Debate, pp. 38-80, Cambridge University Press, 2012. ISBN 9780521875042doi:10.1017/CBO9780511842368.004
27. Powell, James Lawrence (2015). Four Revolutions in the Earth Sciences: From Heresy to Truth. Columbia University Press. pp. 69–70. ISBN 9780231538459. Retrieved 20 October 2015.
28. Hansen, L. T., Some considerations of, and additions to the Taylor-Wegener hypothesis of continental displacement, Los Angeles, 1946. OCLC 1247437 OCLC
29. R. M. Wood, Coming Apart at the Seams, New Scientist, 24 Jan 1980
30. Plate Tectonics: The Rocky History of an Idea. Wegener›s inability to provide an adequate explanation of the forces responsible for continental drift and the prevailing belief that the earth was solid and immovable resulted in the scientific dismissal of his theories.
31. University of California Museum of Paleontology, Alfred Wegener (1880-1930) (accessed 30 April 2015).
32. Holmes, Arthur (1931). Radioactivity and Earth Movements (PDF). Transactions of the Geological Society of Glasgow. Geological Society of Glasgow: 559–606.
33. Holmes, Arthur (1944). Principles of Physical Geology (1st ed.). Edinburgh: Thomas Nelson & Sons. ISBN 0-17-448020-2.
34. McKie, Robin (28 October 2012). David Attenborough: force of nature. The Observer. London. Retrieved 29 October 2012.
35. See map based on the work of the American paleontologist Charles Schuchert in Wells, H. G.; Huxley, Julian; Wells, G. P. (1931), The Science of life, p. 734
36. Carey, S. W. (1958). Carey, S. W., ed. Continental Drift—A symposium. Hobart: Univ. of Tasmania: 177–363.
37. Scheidegger, Adrian E. (1953), Examination of the physics of theories of orogenesis, GSA Bulletin, **64**: 127–150, Bibcode:1953GSAB...64..127S, doi:10.1130/0016-7606(1953)64[127:EOTPOT]2.0.CO;2

38. Jack Oliver, Who Proved Continental Drift, Dies at 87. The New York Times. 12 January 2011. p. A16. Retrieved 6 June 2013.
39. Isacks, Bryan; Oliver, Jack; Sykes, Lynn R. (15 September 1968). Seismology and the New Global Tectonics. Journal of Geophysical Research. **73** *(18): 5855–5899. Bibcode*:1968JGR....73.5855I. doi:10.1029/JB073i018p05855. Retrieved 6 June 2013.
40. Rejoined continents [This Dynamic Earth, USGS]. USGS.

CHAPTER 6

Mountain Formation Processes

Chapter Contents

Processes within the Earth melt rock, which gathers into large pools. This molten rock is lighter than the solid rock around it, so the pressure at those depths gradually forces it upward. Once it encounters sedimentary rocks near the surface, it begins to push it up ahead of itself. If the molten rock breaks through, it forms a volcano, but if not, it forms a dome mountain.

INTRODUCTION

Mountain formation refers to the geological processes that underlie the formation of mountains. These processes are associated with large-scale movements of the Earth›s crust (plate tectonics). Folding, faulting, volcanic activity, igneous intrusion and metamorphism can all be parts of the orogenic process of mountain building.The understanding of specific landscape features in terms of the underlying tectonic processes is called *tectonic geomorphology*, and the study of geologically young or ongoing processes is called *neotectonics*.

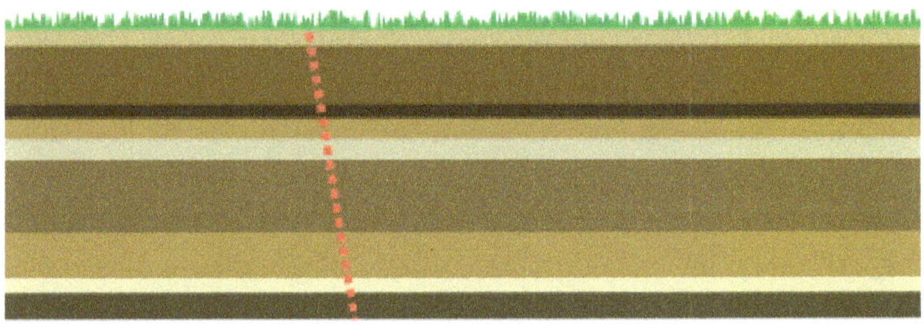

Thrust and reverse fault movement are an important component of mountain formation.

Illustration of mountains that developed on a fold that thrusted.

TYPES OF MOUNTAINS

There are three main types of mountains: volcanic, fold, and block. A more detailed classification useful on a local scale predates plate tectonics and adds to these categories.

Volcanic Mountains

Annotated view includes Ushkovsky, Tolbachik, Bezymianny, Zimina, and Udina stratovolcanoes of Kamchatka, Russia. Oblique view taken on November 12, 2013 from ISS.

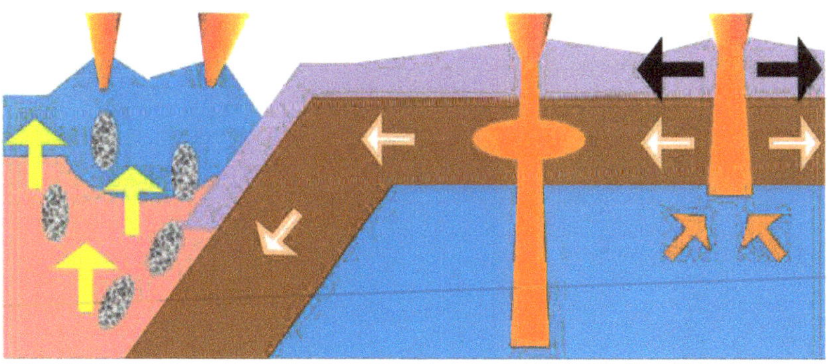

Stratovolcanoes associated with a subduction zone (left) and a spreading ridge volcano (right). A hotspot volcano is center.

Movements of tectonic plates create volcanoes along the plate boundaries, which erupt and form mountains. A *volcanic arc system* is a series of volcanoes that form near a subduction zone where the crust of a sinking oceanic plate melts.

Most volcanoes occur in a band encircling the Pacific Ocean (the Pacific Ring of Fire), and in another that extends from the Mediterranean across Asia to join the Pacific band in the Indonesian Archipelago. The most important types of volcanic mountain are *composite cones* or *stratovolcanoes* (Vesuvius, Kilimanjaro and Mount Fuji are examples) and *shield volcanoes* (such as Mauna Loa on Hawaii, a hotspot volcano).

A shield volcano has a gently sloping cone due to the low viscosity of the emitted material, primarily basalt. Mauna Loa is the classic example, with a slope of 4°-6°. (The relation between slope and viscosity falls under the topic of angle of repose.) The composite volcano or stratovolcano has a more steeply rising cone (33°-40°), due to the higher viscosity of the emitted material, and eruptions are more violent and less frequent than for shield volcanoes. Besides the examples already mentioned are Mount Shasta, Mount Hood and Mount Rainier.

Fold Mountains

Zard-Kuh, a fold mountain in the central Zagros range of Iran.

When plates collide or undergo subduction (that is – ride one over another), the plates tend to buckle and fold, forming mountains. Most of the major continental mountain ranges are associated with thrusting and folding or orogenesis. Examples are the Jura and the Zagros mountains.

Block Mountains

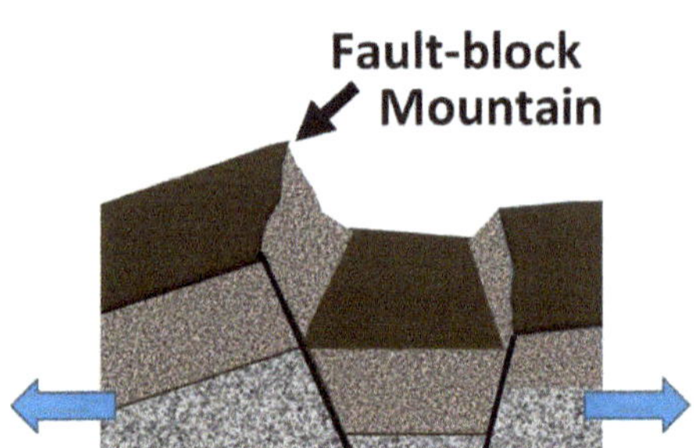

Fault-block mountain of tilted type.

Sierra Nevada Mountains (formed by delamination) as seen from the International Space Station.

When a fault block is raised or tilted, block mountains can result. Higher blocks are called *horsts* and troughs are called *grabens*. A spreading apart of the surface causes tensional forces. When the tensional forces are strong enough to cause a plate to split apart, it does so such that a center block drops down relative to its flanking blocks.

An example is the Sierra Nevada Range, where delamination created a block 650 km long and 80 km wide that consists of many individual portions tipped gently west, with east facing slips rising abruptly to produce the highest mountain front in the continental United States.

Uplifted Passive Margins

Unlike orogenic mountains there is no geophysical model that explains elevated passive continental margins such as the Scandinavian Mountains, Eastern Greenland, the Brazilian Highlands or Australia's Great Dividing Range.

MODELS

Hotspot Volcanoes

Hotspots are supplied by a magma source in the Earth's mantle called a mantle plume. Although originally attributed to a melting of subducted oceanic crust, recent evidence belies this connection. The mechanism for plume formation remains a research topic.

Fault blocks

Several movements of the earth's crust that lead to mountains are associated with faults. These movements actually are amenable to analysis that can predict, for example, the height of a raised block and the width of an intervening rift between blocks using the rheology of the layers and the forces of isostasy. Early bent plate models predicting fractures and fault movements have evolved into today›s kinematic and flexural models.

OROGENY

Orogeny refers to forces and events leading to a large structural deformation of the Earth's lithosphere (crust and uppermost mantle) due to the interaction between tectonic plates. **Orogens** or **orogenic belts** develop when a continental plate is crumpled and is pushed upwards to form mountain ranges, and involve a great range of geological processes collectively called **orogenesis**.

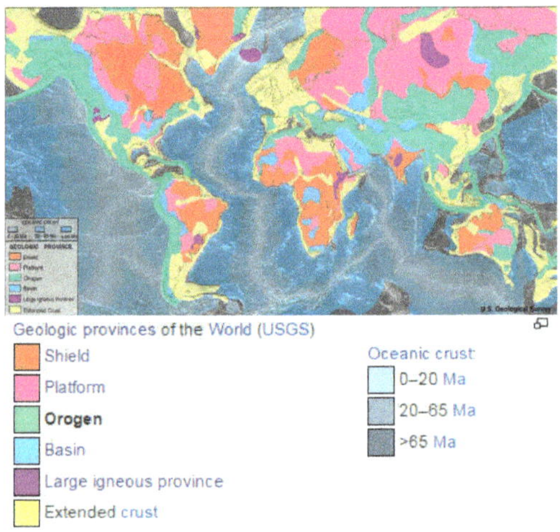

Orogeny is the primary mechanism by which mountains are built on continents. The word "orogeny" comes from Ancient Greek. Though it was used before him, the term was employed by the American geologist G.K. Gilbert in 1890 to describe the process of mountain building as distinguished from epeirogeny.

Physiography

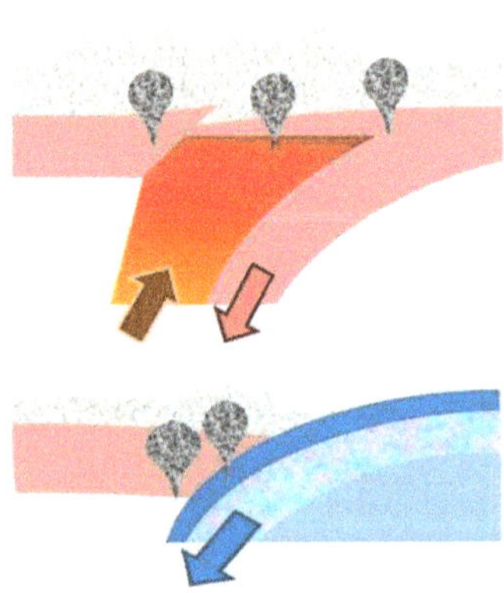

Two processes that can contribute to an orogen. Top: delamination by intrusion of hot asthenosphere; Bottom: Subduction of ocean crust. The two processes lead to differently located granites (bubbles in diagram), providing evidence as to which process actually occurred.

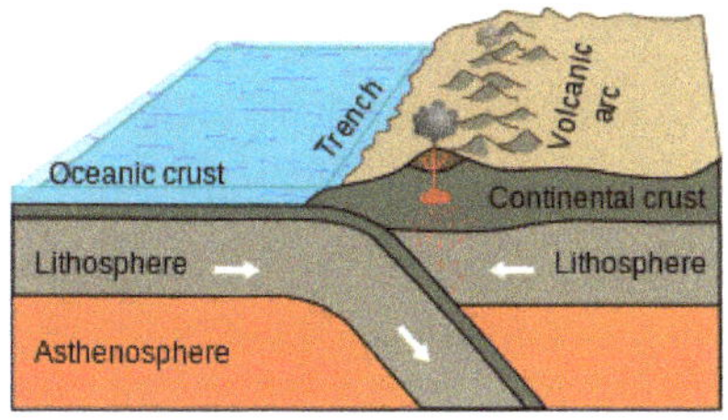

Subduction of an oceanic plate by a continental plate to form a noncollisional orogen. (example: the Andes).

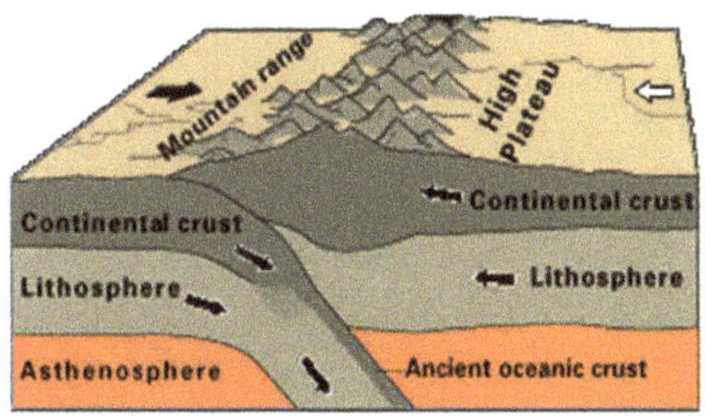

Continental collision of two continental plates to form a collisional orogen. However, usually no continental crust is subducted, only uplifted. (example: the Alps).

Formation of an orogen is accomplished in part by the tectonic processes of subduction (where a continent rides forcefully over an oceanic plate (noncollisional orogens)) or convergence of two or more continents (collisional orogens).

Orogeny usually produces long arcuate (from Latin *arcuare*, "to bend like a bow") structures, known as *orogenic belts*. Generally, orogenic belts consist of long parallel strips of rock exhibiting similar characteristics along the length of the belt. Orogenic belts are associated with subduction zones, which consume crust, produce volcanoes, and build island arcs. Geologists attribute the arcuate structure to the rigidity of the descending plate, and island arc cusps relate to tears in the descending lithosphere. These island arcs may be added to a continent during an orogenic event.

The processes of orogeny can take tens of millions of years and build mountains from plains or from the seabed. The topographic height of orogenic mountains is related to the principle of isostasy, that is, a balance of the downward gravitational force upon an upthrust mountain range (composed of light, continental crust material) and the buoyant upward forces exerted by the dense underlying mantle.

Frequently, rock formations that undergo orogeny are severely deformed and undergo metamorphism. Orogenic processes may push deeply buried rocks to the surface. Sea-bottom and near-shore material may cover some or all of the orogenic area. If the orogeny is due to two continents colliding, very high mountains can result (see Himalayas).

An orogenic event may be studied: (a) as a tectonic structural event, (b) as a geographical event, and (c) as a chronological event.

Orogenic events:

- cause distinctive structural phenomena related to tectonic activity
- affect rocks and crust in particular regions, and
- happen within a specific period

Orogen (or "orogenic system")

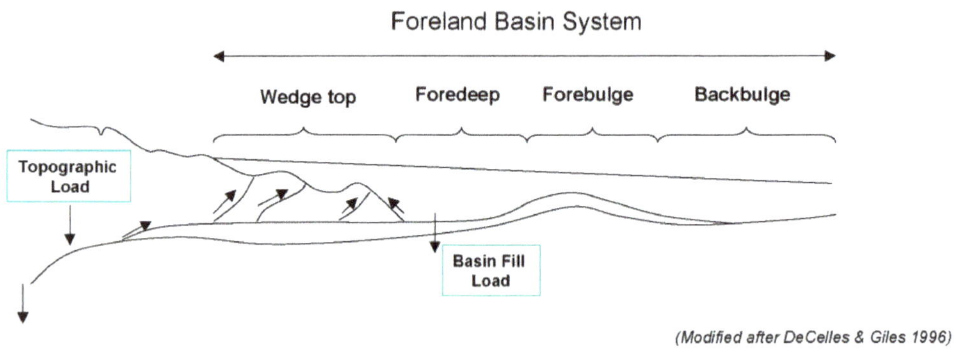

The Foreland Basin System.

An orogeny produces an *orogen*, or (mountain) range-*foreland basin* system.

The *foreland basin* forms ahead of the orogen due mainly to loading and resulting flexure of the lithosphere by the developing mountain belt. A typical foreland basin is subdivided into a wedge-top basin above the active orogenic wedge, the foredeep immediately beyond the active front, a forebulge high of flexural origin and a back-bulge area beyond, although not all of these are present in all foreland-basin systems. The basin migrates with the orogenic front and early deposited foreland basin sediments become progressively involved in folding and thrusting. Sediments deposited in the foreland basin are mainly derived from the erosion of the actively uplifting rocks of the mountain range, although some sediments derive from the foreland. The fill of many such basins shows a change in time from deepwater marine (*flysch*-style) through shallow water to continental (*molasse*-style) sediments.

Orogenic Cycle

Although orogeny involves plate tectonics, the tectonic forces result in a variety of associated phenomena, including magmatism, metamorphism, crustal melting, and crustal thickening. What exactly happens in a specific orogen depends upon the strength and rheology of the continental lithosphere, and how these properties change during orogenesis.

In addition to orogeny, the orogen (once formed) is subject to other processes, such as sedimentation and erosion. The sequence of repeated cycles of sedimentation, deposition and erosion, followed by burial and metamorphism, and then by formation of granitic batholiths and tectonic uplift to form mountain chains, is called the *orogenic cycle*. For example, the *Caledonian Orogeny* refers to the Silurian and Devonian events that resulted from the collision of Laurentia with Eastern Avalonia and other former fragments of Gondwana. The *Caledonian Orogen* resulted from these events and various others that are part of its peculiar orogenic cycle.

In summary, an orogeny is a long-lived deformational episode during which many geological phenomena play a role. The orogeny of an orogen is only part of the orogen›s orogenic cycle.

Erosion

Erosion represents a subsequent phase of the orogenic cycle. Erosion inevitably removes much of the mountains, exposing the core or *mountain roots* (metamorphic rocks brought to the surface from a depth of several kilometres). Isostatic movements may help such exhumation by balancing out the buoyancy of the evolving orogen. Scholars debate about the extent to which erosion modifies the patterns of tectonic deformation (see erosion and tectonics). Thus, the final form of the majority of old orogenic belts is a long arcuate strip of crystalline metamorphic rocks sequentially below younger sediments which are thrust atop them and which dip away from the orogenic core.

An orogen may be almost completely eroded away, and only recognizable by studying (old) rocks that bear traces of orogenesis. Orogens are usually long, thin, arcuate tracts of rock that have a pronounced linear structure resulting in terranes or blocks of deformed rocks, separated generally by suture zones or dippingthrust faults. These thrust faults carry relatively thin slices of rock (which are called nappes or thrust sheets, and differ from tectonic plates) from the core of the shortening orogen out toward the margins, and are intimately associated with folds and the development of metamorphism.

Biology

In the 1950s and 1960s the study of orogeny, coupled with biogeography (the study of the distribution and evolution of flora and fauna), geography and mid ocean ridges, contributed greatly to the theory of plate tectonics. Even at a very early stage, life played a significant role in the continued existence of oceans, by affecting the composition of the atmosphere. The existence of oceans is critical to sea-floor spreading and subduction.

Relationship to Mountain Building

Mountain formation occurs through a number of mechanisms.

Mountain complexes result from irregular successions of tectonic responses due to sea-floor spreading, shifting lithosphere plates, transform faults, and colliding, coupled and uncoupled continental margins.

—Peter J Coney

Large modern orogenies often lie on the margins of present-day continents; the Alleghenian (Appalachian), Laramide, and Andean orogenies exemplify this in the Americas. Older inactive orogenies, such as the Algoman, Penokean and Antler, are represented by deformed rocks and sedimentary basins further inland.

An example of thin-skinned deformation (thrust faulting) of the Sevier Orogeny in Montana. Note the white Madison Limestone repeated, with one example in the foreground (that pinches out with distance) and another to the upper right corner and top of the picture.

Areas that are rifting apart, such as mid-ocean ridges and the East African Rift, have mountains due to thermal buoyancy related to the hot mantle underneath them; this thermal buoyancy is known as dynamic topography. In strike-slip orogens, such as the San Andreas Fault, restraining bends result in regions of localized crustal shortening and mountain building without a plate-margin-wide orogeny. Hotspot volcanism results in the formation of isolated mountains and mountain chains that are not necessarily on tectonic-plate boundaries.

Regions can also experience uplift as a result of delamination of the lithosphere, in which an unstable portion of cold lithospheric root drips down into the mantle, decreasing the density of the lithosphere and causing buoyant uplift. An example is the Sierra Nevada in California. This range of fault-block mountains experienced renewed uplift after a delamination of the lithosphere beneath them.

Finally, uplift and erosion related to epeirogenesis (large-scale vertical motions of portions of continents without much associated folding, metamorphism, or deformation) can create local topographic highs.

Mount Rundle, Banff, Alberta.

Mount Rundle on the TransCanada Highway between Banff and Canmore provides a classic example of a mountain cut in dipping-layered rocks. Millions of years ago a collision caused an orogeny forcing horizontal layers of an ancient ocean crust to be thrust up at an angle of 50–60°. That left Rundle with one sweeping, tree-lined smooth face, and one sharp, steep face where the edge of the uplifted layers are exposed.

The collision causing the Columbia Orogeny occurred about 175 million years ago, and as the shock wave moved eastward, it forced huge masses of rock to crack and slide up over its neighbours. This is known as thrust faulting and was instrumental in the formation of the Rockies. The shock wave began piling up the western ranges, and then the main ranges, around 120 million years ago.

—Mountains in Nature

History of the Concept

Before the development of geologic concepts during the 19th century, the presence of marine fossils in mountains was explained in Christian contexts as a result of the Biblical Deluge. This was an extension of Neoplatonic thought, which influenced early Christian writers.

The 13th-century Dominican scholar Albert the Great posited that, as erosion was known to occur, there must be some process whereby new mountains and other land-forms were thrust up, or else there would eventually be no land; he suggested that marine fossils in mountainsides must once have been at the sea-floor. Orogeny was used by Amanz Gressly (1840) and Jules Thurmann (1854) as *orogenic* in terms of the creation of mountain elevations, as the term *mountain building* was still used to describe the processes. Elie de Beaumont (1852) used the evocative "Jaws of a Vise" theory to explain orogeny, but was more concerned with the height rather than the implicit structures created by and contained in orogenic belts. His theory essentially held that mountains were created by the squeezing of certain rocks. Eduard Suess (1875) recognised the importance of horizontal movement of rocks. The concept of a *precursor geosyncline* or initial downward warping of the solid earth (Hall, 1859) prompted James Dwight Dana (1873) to include the concept of *compression* in the theories surrounding mountain-building. With hindsight, we can discount Dana's conjecture that this contraction was due to the cooling of the Earth (aka the cooling Earth theory). The cooling Earth theory was the chief paradigm for most geologists until the 1960s. It was, in the context of orogeny, fiercely contested by proponents of vertical movements in the crust (similar to tephrotectonics), or convection within the asthenosphere or mantle.

Gustav Steinmann (1906) recognised different classes of orogenic belts, including the *Alpine type orogenic belt*, typified by a flysch and molasse geometry to the sediments; ophiolite sequences, tholeiitic basalts, and a nappe style fold structure.

In terms of recognising orogeny as an *event*, Leopold von Buch (1855) recognised that orogenies could be placed in time by bracketing between the youngest deformed rock and the oldest undeformed rock, a principle which is still in use today, though commonly investigated by geochronology using radiometric dating.

H.J. Zwart (1967) drew attention to the metamorphic differences in orogenic belts, proposing three types, modified by W. S. Pitcher in 1979 and further modified as:

- Hercynotype (back-arc basin type);
  - Shallow, low-pressure metamorphism; thin metamorphic zones
  - Metamorphism dependent on increase in temperature
  - Abundant granite and migmatite
  - Few ophiolites, ultramafic rocks virtually absent
  - very wide orogen with small and slow uplift
  - nappe structures rare
- Alpinotype (ocean trench style);
  - deep, high pressure, thick metamorphic zones
  - metamorphism of many facies, dependent on decrease in pressure
  - few granites or migmatites
  - abundant ophiolites with ultramafic rocks
  - Relatively narrow orogen with large and rapid uplift
  - Nappe structures predominant
- Cordilleran (arc) type;
  - dominated by calc-alkaline igneous rocks, andesites, granite batholiths
  - general lack of migmatites, low geothermal gradient
  - lack of ophiolite and abyssal sedimentary rocks (black shale, chert, etcetera)

- low-pressure metamorphism, moderate uplift
- lack of nappes

The advent of plate tectonics has explained the vast majority of orogenic belts and their features. The cooling earth theory (principally advanced by Descartes) is dispensed with, and tephrotectonic style vertical movements have been explained primarily by the process of isostasy.

Some oddities exist, where simple collisional tectonics are modified in a transform plate boundary, such as in New Zealand, or where island arc orogenies, for instance in New Guinea occur away from a continental backstop. Further complications such as Proterozoic continent-continent collisional orogens, explicitly the Musgrave Block in Australia, previously inexplicable are being brought to light with the advent of seismic imaging techniques which can resolve the deep crust structure of orogenic belts.

PLATE RECONSTRUCTION

Plate reconstruction is the process of reconstructing the positions of tectonic plates relative to each other (relative motion) or to other reference frames, such as the earth›s magnetic field or groups of hotspots, in the geological past. This helps determine the shape and make-up of ancient supercontinents and provides a basis for paleogeographic reconstructions.

Defining Plate Boundaries

An important part of reconstructing past plate configurations is to define the edges of areas of the lithosphere that have acted independently at some time in the past.

Present Plate Boundaries

Most present plate boundaries are easily identifiable from the pattern of recent seismicity. This is now backed up by the use of GPS data, to confirm the presence of significant relative movement between plates.

Past Plate Boundaries

Identifying past (but now inactive) plate boundaries within current plates is generally based on evidence for an ocean that has now closed up. The line where the ocean used to be is normally marked by pieces of the crust from that ocean, included in the collision zone, known as ophiolites. The line across which two plates became joined to form a single larger plate, is known as a suture.

In many orogenic belts, the collision is not just between two plates, but involves the sequential accretion of smaller terranes. Terranes are smaller pieces of continental crust that have been caught up in an orogeny, such as continental fragments or island arcs.

Reference Frames

Plate motions, both those observable now and in the past, are referred ideally to a reference frame that allows other plate motions to be calculated. For example, a central plate, such as the African plate, may have the motions of adjacent plates referred to it. By composition of reconstructions, additional plates can be reconstructed to the central plate. In turn, the reference plate may be reconstructed, together with the other plates, to another reference frame, such as the earth's magnetic field, as determined from paleomagnetic measurements of rocks of known age. A global hotspot reference frame has been postulated (see, e.g., W. Jason Morgan) but there is now evidence that not all hotspots are necessarily fixed in their locations relative to one another or the earth›s spin axis. However, there are groups of such hotspots that appear to be fixed within the constraints of available data, within particular mesoplates.

Euler poles

The movement of a rigid body, such as a plate, on the surface of a sphere can be described as rotation about a fixed axis (relative to the chosen reference frame). This pole of rotation is known as an Euler pole. The movement of a plate is completely specified in terms of its Euler pole and the angular rate of rotation about the pole. Euler poles defined for current plate motions can be used to reconstruct plates in the recent past (few million years). At earlier stages of earth›s history, new Euler poles need to be defined. It was suggested recently that paleomagnetic Euler poles derived from apparent polar wander path could potentially extend the upper age limit of marine geophysical observations and hot spot traces.

Estimating Past Plate Motions

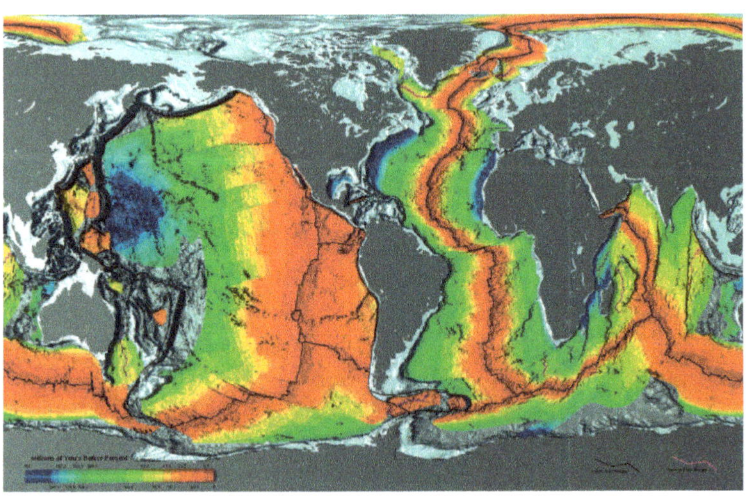

Ages of oceanic lithosphere.

In order to move plates backward in time it is necessary to provide information on either relative or absolute positions of the plates being reconstructed such that an Euler pole can be calculated. These are quantitative methods of reconstruction.

Geometric Matching of Continental Borders

Certain fits between continents particularly that between South America and Africa, were known long before the development of a theory that could adequately explain them. The reconstruction before Atlantic rifting by Bullard based on a least-squares fitting at the 500 fathom contour still provides the best match to paleomagnetic pole data for the two sides from the middle of Paleozoic to Late Triassic.

Plate Motion from Magnetic Stripes

Plate reconstructions in the recent geological past mainly use the pattern of magnetic stripes in oceanic crust to remove the effects of seafloor spreading. The individual stripes are dated from magnetostratigraphy so that their time of formation is known. Each stripe (and its mirror image) represents a plate boundary at a particular time in the past, allowing the two plates to be repositioned relative to one another. The oldest oceanic crust is of Jurassic age, providing a lower age limit of about 175 Ma for the use of such data. Reconstructions derived in this way are only relative.

Paleomagnetic Pole Data

Sampling

Paleomagnetic pole data is obtained by taking oriented samples of rock. Good quality poles have been recovered from many different rock types. In igneous rocks, the magnetic minerals have cooled through the curie point, typically providing good quality poles. Metamorphic rocks may contain new magnetic minerals but are not normally used to obtain poles due to their typically high magnetic anisotropy. Sedimentary rocks generally have a complex set of magnetic components such as derived clasts and mineral grains with their own inherited magnetization, primary magnetization acquired at the time of sedimentation and secondary magnetization related to subsequent diagenesis and authigenic mineral growth. Progressive demagnetization is used to discriminate between the components to identify the primary magnetization. The poles shown by the primary magnetization represent a record of their position relative to the magnetic pole at the time when the rock was formed. Suitable rocks of the right age are not present everywhere, so some pieces of the puzzle remain very unconstrained.

Good quality magnetic poles are added to the *Global Paleomagnetic Database*, which is accessible from the World Data Center A in the USA at Boulder, Colorado; and from the Norwegian Geological Survey in Trondheim.

Apparent polar Wander Paths

Poles from different ages in a single area can be used to construct an apparent polar wander (APW) path. If paths from adjacent crustal fragments are identical, this is taken to indicate that there has been no relative movement between them during the period covered by the path. Divergence of APW paths indicates that the areas in question have acted independently in the past with the point of divergence marking the time at which they became joined.

Observed paleomagnetic poles show only the approximate latitude at the time but provide no constraint on the longitude. For this reason older plate reconstructions have much larger uncertainties associated with them. Paleomagnetic Euler rotations (rotation poles and angles) and the mantle structures, however, could provide clues to this longstanding challenge.

Hotspot Tracks

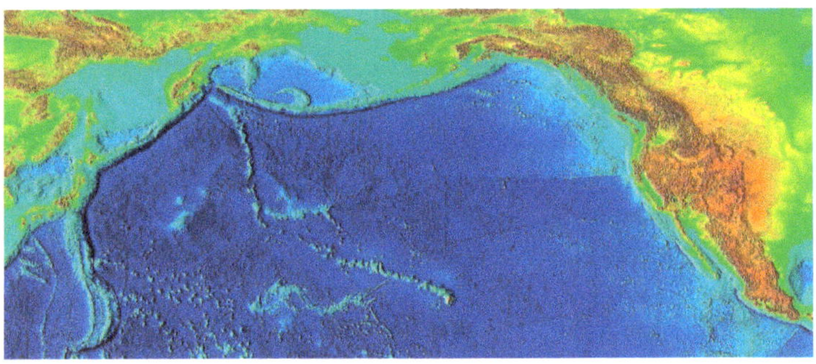

The Hawaiian-Emperor seamount chain.

The presence of chains of volcanic islands and seamounts interpreted to have formed from fixed hotspots allows the plate on which they sit to be progressively restored so that a seamount is moved back over the hotspot at its time of formation. This method can be used back to the Early Cretaceous, the age of the oldest evidence for hotspot activity. This method gives an absolute reconstruction of both latitude and longitude, although before about 90 Ma there is evidence of relative motion between hotspot groups.

Slab Constraints

Once oceanic plates subduct in the lower mantle (slabs), they are assumed to sink in a near-vertical manner. With the help of seismic wave tomography, this can be used to constrain plate reconstructions at first order back to the Permian.

Apparent Polar Wander Paths Geometric Parameterizations

Paleomagnetic Euler poles derived by geometrizing apparent polar wander paths (APWPs) potentially allows constraining paleolongitudes from paleomagnetic data. This method could extend absolute plate motion reconstructions deeply into the geologic history as long as there are reliable APWPs.

Other evidence for Past Plate Configurations

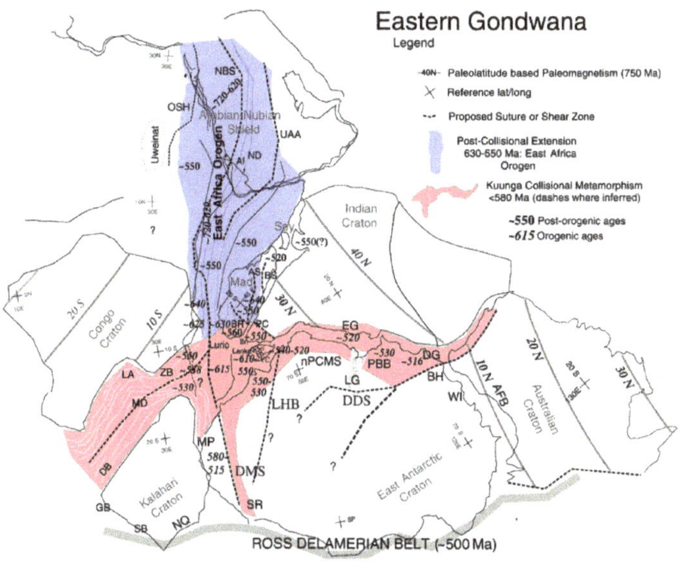

Reconstruction of eastern Gondwana showing position of orogenic belts.

Some plate reconstructions are supported by other geological evidence, such as the distribution of sedimentary rock types, the position of orogenic belts and faunal provinces shown by particular fossils. These are semi-quantitative methods of reconstruction.

Sedimentary Rock Types

Some types of sedimentary rock are restricted to certain latitudinal belts. Glacial deposits for instance are generally confined to high latitudes, whereas evaporites are generally formed in the tropics.

Faunal Provinces

Oceans between continents provide barriers to plant and animal migration. Areas that have become separated tend to develop their own fauna and flora. This is particularly the case for plants and land animals but is also true for shallow water marine species, such as trilobites and brachiopods, although their planktonic larvae mean that they were able to migrate over smaller deep water areas. As oceans narrow before a collision occurs, the faunas start to become mixed again, providing supporting evidence for the closure and its timing.

Orogenic Belts

When supercontinents break up, older linear geological structures such as orogenic belts may be split between the resulting fragments. When a reconstruction effectively joins up orogenic belts of the same age of formation, this provides further support for the reconstruction's validity.

REFERENCES

1. Steven M. Stanley (2004). "Mountain building". Earth system history (2nd ed.). Macmillan. p. 207. ISBN 0-7167-3907-0.
2. Robert J. Twiss; Eldridge M. Moores (1992). Plate tectonic models of orogenic core zones. Structural Geology (2nd ed.). Macmillan. p. 493. ISBN 0-7167-2252-6.
3. Kurt Stüwe (2007). §4.5 Geomorphology. Geodynamics of the lithosphere: an introduction (2nd ed.). Springer. p. 178. ISBN 3-540-71236-4.
4. Mountain building. Science matters: earth and beyond; module 4. Pearson South Africa. 2002. p. 75. ISBN 0-7986-6059-7.
5. Andrew Goudie (2004). Encyclopedia of geomorphology; Volume 2. Routledge. p. 701. ISBN 0-415-32738-5.
6. NASA - Activity at Kliuchevskoi
7. Victor Schmidt; William Harbert (2003). Planet Earth and the New Geoscience (4th ed.). Kendall Hunt. pp. 46–47. ISBN 0-7872-9355-5.
8. Stephen D Butz (2004). Chapter 8: Plate tectonics. Science of Earth Systems. Thompson/Delmar Learning. p. 136. ISBN 0-7668-3391-7.
9. John Gerrard (1990). Types of volcano. Mountain environments: an examination of the physical geography of mountains. MIT Press. p. 194. ISBN 0-262-07128-2.
10. Robert Wayne Decker; Barbara Decker (2005). Chapter 8: Hot spots. Volcanoes (4th ed.). Macmillan. p. 113 ff. ISBN 0-7167-8929-9.
11. Arthur Holmes; Donald Duff (2004). Holmes Principles of Physical Geology (4th ed.). Taylor & Francis. p. 209. ISBN 0-7487-4381-2.
12. Transactions of the American Society of Civil Engineers, Volume 39. American Society of Civil Engineers. 1898. p. 62.
13. James Shipman; Jerry D. Wilson; Aaron Todd (2007). Minerals, rocks and volcanoes. An Introduction to Physical Science (12th ed.). Cengage Learning. p. 650. ISBN 0-618-93596-7.
14. Michael P Searle (2007). Diagnostic features and processes in the construction and evolution of Oman-, Zagros-, Himalayan-, Karakoram-, and Tibetan type orogenic belts. In Robert D. Hatcher Jr.; MP Carlson; JH McBride & JR Martinez Catalán. 4-D framework of continental crust. Geological Society of America. p. 41 ff. ISBN 0-8137-1200-9.
15. Chris C. Park (2001). Figure 6.11. The environment: principles and applications (2nd ed.). Routledge. p. 160. ISBN 0-415-21770-9.
16. Scott Ryan (2006). Figure 13-1. CliffsQuickReview Earth Science. Wiley. ISBN 0-471-78937-2.
17. John Gerrard (1990-04-12). Reference cited. p. 9. ISBN 0-262-07128-2.

18. Lee, C.-T.; Yin, Q; Rudnick, RL; Chesley, JT; Jacobsen, SB (2000). Osmium Isotopic Evidence for Mesozoic Removal of Lithospheric Mantle Beneath the Sierra Nevada, California (PDF). Science. **289** *(5486): 1912–6. Bibcode:2000Sci...289.1912L. doi:10.1126/science.289.5486.1912. PMID 10988067.*

19. Bonow, Johan M. (2009). atlantens kustberg och högslätter – gamla eller unga? (PDF). www.geografitorget.se (in Swedish). Geografilärarnas Riksförening.

20. Green, Paul F.; Lidmar-Bergström, Karna; Japsen, Peter; Bonow, Johan M.; Chalmers, James A. (2013). Stratigraphic landscape analysis, thermochronology and the episodic development of elevated, passive continental margins. Geological Survey of Denmark and Greenland Bulletin. **30**: 18. Retrieved 30 April 2015.

21. Y Niu & MJ O›Hara (2004). Chapter 7: Mantle plumes are NOT from ancient oceanic crust. In Roger Hékinian; Peter Stoffers & Jean-Louis Cheminée. Oceanic hotspots: intraplate submarine magmatism and tectonism. Springer. p. 239 ff. ISBN 3-540-40859-2.

22. AB Watts (2001). §7.2 Extensional tectonics and rifting. Isostasy and flexure of the lithosphere. Cambridge University Press. p. 295. ISBN 0-521-00600-7.

23. GD Karner & NW Driscoll (1999). Style, timing and distribution of tectonic deformation across the Exmouth Plateau, northwest Australia, determined from stratal architecture and quantitative basin modelling. In Conall Mac Niocaill & Paul Desmond Ryan. Continental tectonics. Geological society. p. 280. ISBN 1-86239-051-7.

24. Tony Waltham (2009). Foundations of Engineering Geology (3rd ed.). Taylor & Francis. p. 20. ISBN 0-415-46959-7.

25. Philip Kearey; Keith A. Klepeis; Frederick J. Vine (2009). Chapter 10: Orogenic belts. Global Tectonics (3rd ed.). Wiley-Blackwell. p. 287. ISBN 1-4051-0777-4.

26. Chambers 21st Century Dictionary. Allied Publishers. 1999. p. 972. ISBN 978-0550106254. Retrieved 27 June 2012.

27. Friedman G.M. (1994). Pangean Orogenic and Epeirogenic Uplifts and Their Possible Climatic Significance. In Klein G.O. Pangea: Paleoclimate, Tectonics, and Sedimentation During Accretion, Zenith, and Breakup of a Supercontinent. Geological Society of America Special Paper. **288**. p. 160. ISBN 9780813722887.

28. N. H. Woodcock; Robin A. Strachan (2000). Chapter 12: The Caledonian Orogeny: a multiple plate collision. Geological History of Britain and Ireland. Wiley-Blackwell. p. 202, Figure 12.11. ISBN 0-632-03656-7.

29. Frank Press (2003). Understanding Earth (4th ed.). Macmillan. pp. 468–469. ISBN 0-7167-9617-1.

30. Gerald Schubert; Donald Lawson Turcotte; Peter Olson (2001). §2.5.4 Why are island arcs arcs?. Mantle Convection in the Earth and Planets. Cambridge University Press. pp. 35–36. ISBN 0-521-79836-1.

31. PA Allen (1997). Isostasy in zones of convergence. Earth Surface Processes. Wiley-Blackwell. p. 36 ff. ISBN 0-632-03507-2.

32. Gerard V. Middleton; Peter R. Wilcock (1994). §5.5 Isostasy. Mechanics in the Earth and Environmental Sciences (2nd ed.). Cambridge University Press. p. 170. ISBN 0-521-44669-4.

33. DeCelles P.G. & Giles K.A. (1996). Foreland basin systems (PDF). Basin Research. **8** *(2): 105–123. doi:10.1046/j.1365-2117.1996.01491.x.*

34. David Johnson (2004). The orogenic cycle. The geology of Australia. Cambridge University Press. p. 48 ff. ISBN 0-521-84121-6.

35. In other words, orogeny is only a phase in the existence of an orogen. Five characteristics of the orogenic cycle are listed by Robert J. Twiss; Eldridge M. Moores (1992). Plate tectonic models of orogenic core zones. Structural Geology (2nd ed.). Macmillan. p. 493. ISBN 0-7167-2252-6.

36. N. H. Woodcock; Robin A. Strachan (2000). Chapter 12: The Caledonian Orogeny: A Multiple Plate Collision. cited work. p. 187 ff. ISBN 0-632-03656-7.

37. Olivier Merle (1998). §1.1 Nappes, overthrusts and fold-nappes. Emplacement Mechanisms of Nappes and Thrust Sheets. Petrology and Structural Geology. **9**. Springer. p. 1 ff. ISBN 0-7923-4879-6.

38. For example, see Patrick L Osborne (2000). Tropical Ecosystems and Ecological Concepts. Cambridge University Press. p. 11. ISBN 0-521-64523-9. Continental drift and plate tectonics help to explain both the similarities and the differences in the distribution of plants and animals over the continents and John C Briggs (1 987). Biogeography and Plate Tectonics. Elsevier. p. 131. ISBN 0-444-42743-0. It will not be possible to construct a thorough account of the history of the southern hemisphere without the evidence from both the biological and the earth sciences Check date values in: |date= (help)

39. Paul D. Lowman (2002). Chapter 7: Geology and biology: the influence of life on terrestrial geology. Exploring Space, Exploring Earth: New Understanding of the Earth from Space Research. Cambridge University Press. pp. 286–287. ISBN 0-521-89062-4.

40. Seema Sharma (2005). Atmosphere: origin. Encyclopaedia of Climatology. Anmol Publications PVT. LTD. p. 30 ff. ISBN 81-261-2442-3.

41. Richard J. Huggett (2007). Fundamentals of Geomorphology (2nd ed.). Routledge. p. 104. ISBN 0-415-39084-2.

42. Gerhard Einsele (2000). Sedimentary Basins: Evolution, Facies, and Sediment Budget (2nd ed.). Springer. p. 453. ISBN 3-540-66193-X. Without denudation, even relatively low uplift rates as characteristic of epeirogenetic movements (e.g. 20m/MA) would generate highly elevated regions in geological time periods.

43. Ian Douglas; Richard John Huggett; Mike Robinson (2002). Companion Encyclopedia of Geography: The Environment and Humankind. Taylor & Francis. p. 33. ISBN 0-415-27750-7.

44. Peter J Coney (1970). The Geotectonic Cycle and the New Global Tectonics. Geological Society of America Bulletin. **81** *(3): 739–748. Bibcode:1970GSAB...81..739C. doi:10.1130/0016-7606(1970)81[739:TGCATN]2.0.CO;2.*

45. Lee, C.-T.; Yin, Q; Rudnick, RL; Chesley, JT; Jacobsen, SB (2000). Osmium Isotopic Evidence for Mesozoic Removal of Lithospheric Mantle Beneath the Sierra Nevada, California (PDF). Science. **289** *(5486): 1912–6. Bibcode:2000Sci...289.1912L. doi:10.1126/science.289.5486.1912. PMID 10988067.*

46. John Gerrard (1990). Mountain Environments: An Examination of the Physical Geography of Mountains. MIT Press. p. 9. ISBN 0-262-07128-2.

47. Manley, Curtis R.; Glazner, Allen F.; Farmer, G. Lang (2000). Timing of Volcanism in the Sierra Nevada of California: Evidence for Pliocene Delamination of the Batholithic Root?. Geology. **28** *(9): 811. Bibcode:2000Geo....28..811M. doi:10.1130/0091-7613(2000)28<811:TOVITS>2.0.CO;2.*

48. Arthur Holmes; Doris L. Holmes (2004). Holmes Principles of Physical Geology (4th ed.). Taylor & Francis. p. 92. ISBN 0-7487-4381-2.

49. The Formation of the Rocky Mountains, Mountains in Nature, n.d., retrieved 29 January 2014

50. Zwart, H. J., 1967. The duality of orogenicb elts. Geol. Mijnbouw. 46, 283-309 (referred in Pitcher 1979)

51. W.S. Pitcher, "The nature, ascent and emplacement of granitic magmas", Journal of the Geological Society 1979; v. 136; p. 627-662

52. de Beaumont, J.-B., Élie (1852 (English synopsis in Dennis (1982))). Notice sur les Systèmes de Montagnes (Note on Mountain Systems). Bertrand, Paris. pp. 1543 pp. Check date values in: |date= (help)

53. Condie, K.C. (1997). Plate tectonics and crustal evolution (4th ed.). Butterworth-Heinemann. p. 282. ISBN 978-0-7506-3386-4. Retrieved 2010-02-21.

54. Lliboutry, L. (2000). Quantitative geophysics and geology. Springer. p. 480. ISBN 978-1-85233-115-3. Retrieved 2010-02-22.

55. Kearey, P.; Klepeis K.A. & Vine F.J. (2009). Global tectonics (3rd ed.). Wiley-Blackwell. p. 482. ISBN 978-1-4051-0777-8.

56. Pilger, R.H. (2003). Geokinematics: prelude to geodynamics. Springer. p. 338. Retrieved 2010-02-21.

57. Wu, L.; Kravchinsky V.A. Derivation of paleo-longitude from the geometric parametrization of apparent polar wander path: implication for absolute plate motion reconstruction. Retrieved 10 February 2014.

58. Torsvik, T.H. Reconstruction Methods. Retrieved 21 February 2010.

59. Merrill, R.T.; McElhinny M.W. The earth›s magnetic field: its history, origin, and planetary perspective. International geophysics series. **32**. Academic Press. p. 401. ISBN 978-0-12-491240-3. Retrieved 2010-02-21.

60. National Geophysics Data Center (2010). IAGA Paleomagnetic Databases. Retrieved 21 February 2010.

61. Geological Survey of Norway (2005). Global paleomagnetic database. Retrieved 21 February 2010.

62. Butler, R.F. (1992). Chapter 10 Applications to paleogeography (PDF). Paleomagnetism:Magnetic domains to geologic terranes. Blackwell. Retrieved 22 February 2010.

63. Torsvik, Trond Helge; Steinberger, Bernhard (December 2006). Fra kontinentaldrift til manteldynamikk [From Continental Drift to Mantle Dynamics]. Geo (in Norwegian). **8**: 20–30. Retrieved 22 June 2010., translation: Torsvik, Trond Helge; Steinberger, Bernhard (2008). "From Continental Drift to Mantle Dynamics" (PDF). In Trond Slagstad; Rolv Dahl Gråsteinen. Geology for Society for 150 years - The Legacy after Kjerulf. **12**. Trondheim: Norges Geologiske Undersokelse. pp. 24–38. Retrieved 18 June 2010

64. van der Meer, D.G.; Spakman W.; van Hinsbergen D.J.J.; Amaru M.L. & Torsvik T.H. (2010). Towards absolute plate motions constrained by lower-mantle slab remnants (PDF). Nature Geoscience. Macmillan Publishers Limited. **3**: 36–40. Bibcode:2010NatGe...3...36V. doi:10.1038/NGEO708. Retrieved 22 November 2011.

65. Scotese, C.R. (2002-04-20). Climate History. Paleomap Project. Retrieved 22 February 2010.

CHAPTER 7

An Overview of Current Tectonic Plates

Chapter Contents

- **Introduction**
- **Current Plates**
- **Ancient Continental Formations**
- **Cimmeria**
- **History of Concept**
- **Tectonic History**
- **Economic Importance**
- **Lhasa Terrane**
- **Agulhas Basin**
- **Farallon Plate**

This is a list of tectonic plates on Earth. Tectonic plates are pieces of Earth›s crust and uppermost mantle, together referred to as the lithosphere. The plates are around 100 km (62 mi) thick and consist of two principal types of material: oceanic crust (also called *sima* from silicon and magnesium) and continental crust(*sial* from silicon and aluminium). The composition of the two types of crust differs markedly, with basaltic rocks (mafic) dominating oceanic crust, while continental crust consists principally of lower density granitic rocks (felsic).

INTRODUCTION

This is a **list of tectonic plates** on Earth. Tectonic plates are pieces of Earth's crust and uppermost mantle, together referred to as the lithosphere. The plates are around 100 km (62 mi) thick and consist of two principal types of material: oceanic crust (also called *sima* from silicon and magnesium) and continental crust (*sial* from silicon and aluminium). The composition of the two types of crust differs markedly, with basaltic rocks ("mafic) dominating oceanic crust, while continental crust consists principally of lower density granitic rocks ("felsic). See also Plate tectonics.

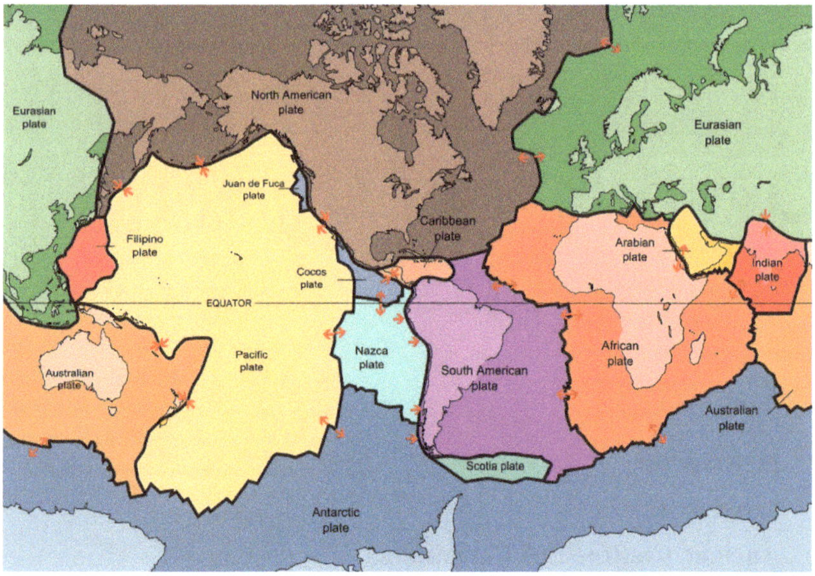

The 15 major plates

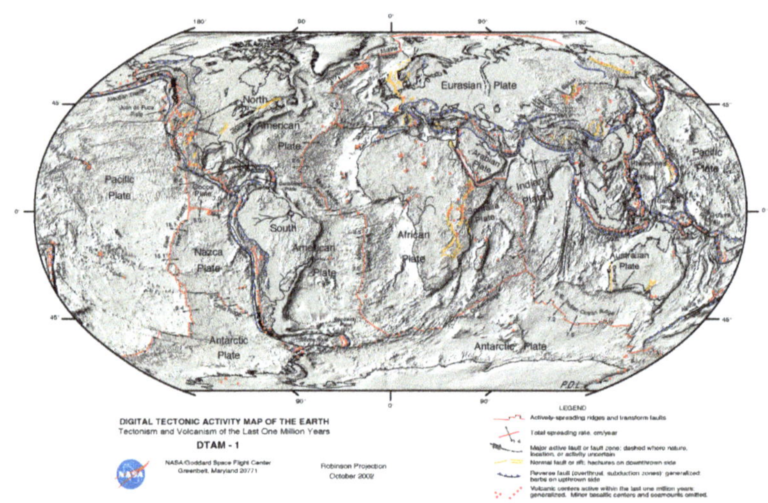

Plate tectonics map from NASA.

CURRENT PLATES

Geologists generally agree that the following tectonic plates currently exist on the Earth's surface with roughly definable boundaries. Tectonic plates are sometimes subdivided into three fairly arbitrary categories: *major* (or *primary*) *plates*, *minor* (or *secondary*) *plates*, and *microplates* (or *tertiary plates*).

Major Plates

These plates comprise the bulk of the continents and the Pacific Ocean. For purposes of this list, a **major plate** is any plate with an area greater than 20 million km^2.

- Pacific Plate – 103,300,000 km^2
- North American Plate – 75,900,000 km^2
- Eurasian Plate – 67,800,000 km^2
- African Plate – 61,300,000 km^2
- Antarctic Plate – 60,900,000 km^2
- Indo-Australian Plate – 58,900,000 km^2 often considered two plates:
    - Australian Plate – 47,000,000 km^2
    - Indian Plate – 11,900,000 km^2
- South American Plate – 43,600,000 km^2

Minor Plates

These smaller plates are often not shown on major plate maps, as the majority do not comprise significant land area. For purposes of this list, a **minor plate** is any plate with an area less than 20 million km^2 but greater than 1 million km^2.

- Somali Plate – 16,700,000 km^2
- Nazca Plate – 15,600,000 km^2
- Philippine Sea Plate – 5,500,000 km^2
- Arabian Plate – 5,000,000 km^2
- Caribbean Plate – 3,300,000 km^2
- Cocos Plate – 2,900,000 km^2
- Caroline Plate – 1,700,000 km^2
- Scotia Plate – 1,600,000 km^2
- Burma Plate – 1,100,000 km^2
- New Hebrides Plate – 1,100,000 km^2

Microplates

These plates are often grouped with an adjacent major plate on a major plate map. For purposes of this list, a microplate is any plate with an area less than 1 million km^2. Some models identify more minor plates within current orogens like the Apulian, Explorer, Gorda, and Philippine Mobile Belt plates. There may or may not be scientific consensus as to whether such plates should be considered distinct portions of the crust, thus new research could change this list.

- African Plate

    - Lwandle Plate
    - Madagascar Plate
    - Rovuma Plate
    - Seychelles microcontinent
- Antarctic Plate
    - Shetland Plate
    - South Sandwich Plate
- Caribbean Plate
    - Panama Plate
    - Gonâve Microplate
- Cocos Plate
    - Rivera Plate
- Eurasian Plate
    - Adriatic Plate (or Apulian Plate)
    - Aegean Sea Plate (or Hellenic Plate)
    - Amurian Plate
    - Anatolian Plate
    - Banda Sea Plate
    - Iberian Plate
    - Iranian Plate
    - Molucca Sea Plate
        - Halmahera Plate
        - Sangihe Plate
    - Okinawa Plate
    - Pelso Plate
    - Sunda Plate
    - Timor Plate
    - Tisza Plate

- Yangtze Plate

- Australian Plate
  - Capricorn Plate
  - Futuna Plate
  - Kermadec Plate
  - Maoke Plate
  - Niuafo›ou Plate
  - Tonga Plate
  - Woodlark Plate
- North American Plate
  - Greenland Plate
  - Okhotsk Plate
  - Juan de Fuca Plate
  - Explorer Plate
  - Gorda Plate
- Pacific Plate
  - Balmoral Reef Plate
  - Bird›s Head Plate
  - Caroline Plate
  - Conway Reef Plate
  - Easter Plate
  - Galapagos Microplate
  - Juan Fernandez Plate
  - Kula Plate
  - Manus Plate
  - New Hebrides Plate
  - North Bismarck Plate

    - North Galápagos Microplate
    - Solomon Sea Plate
    - South Bismarck Plate
- Philippine Sea Plate
    - Mariana Plate
    - Philippine Microplate

South American Plate
- Altiplano Plate
- Falklands Microplate
- North Andes Plate

ANCIENT CONTINENTAL FORMATIONS

In the history of Earth many tectonic plates have come into existence and have over the intervening years either accreted onto other plates to form larger plates, rifted into smaller plates, or have been crushed by or subducted under other plates (or have done all three).

Ancient Supercontinents

A supercontinent is a landmass consisting of multiple continental cores. The following list includes the supercontinents known or speculated to have existed in the Earth's past:

- Columbia
- Euramerica
- Gondwana
- Kenorland
- Laurasia
- Nena
- Pangaea
- Pannotia
- Proto-Laurasia
- Rodinia
- Ur
- Vaalbara

Ancient Plates and Cratons

Not all plate boundaries are easily defined, this is especially true for ancient pieces of crust. The following list of ancient cratons, microplates, plates, shields, terranes, and zones no longer exist as separate plates. *Cratons* are the oldest and most stable parts of the continental lithosphere and *shields* are the exposed area of a craton(s). *Microplates* are tiny tectonic plates, *terranes* are fragments of crustal material formed on one tectonic plate and accreted to crust lying on another plate, and *zones* are bands of similar rocks on a plate formed by terrane accretion or native rock formation. Terranes may or may not have originated as independent microplates since a terrane may not contain the full thickness of the lithosphere.

African Plate

- Atlantica
- Bangweulu Block (Zambia)
- Congo Craton (Angola, Cameroon, Central African Republic, Democratic Republic of Congo, Gabon, Sudan, and Zambia)
- Kaapvaal Craton (South Africa)
- Kalahari Craton (South Africa)
- Saharan Metacraton (Algeria)
- Sebakwe proto-Craton (Zimbabwe)
- Tanzanian Craton (Tanzania)
- West African Craton (Algeria, Benin, Burkina Faso, Côte d›Ivoire, Gambia, Ghana, Guinea, Guinea Bissau, Liberia, Mali, Mauritania, Morocco, Nigeria, Senegal, Sierra Leone, and Togo)
- Zaire Craton (Congo)
- Zimbabwe Craton (Zimbabwe)

Antarctica Plate

- Bellingshausen Plate
- Charcot Plate
- East Antarctic Craton
- Phoenix Plate

Eurasian Plate

- Armorica (France, Germany, Spain and Portugal)
- Avalonia (Canada, Great Britain, and United States)
- Baltic Plate (Russia)

- Belomorian Craton
- Central Iberian Plate (Spain)
- Cimmerian Plate (Anatolia, Iran, Afghanistan, Tibet, Indochina and Malaya)
- East China Craton (China)
- East European Craton (Russia and Scandinavia)
- Fennoscandian Shield (Norway, Sweden, and Finland)
- Junggar Plate (China)
- Hunic plate
- Karelian Craton (Finland and Russia)
- Kazakhstania (Kazakhstan)
- Kola Craton (northwest Russia)
- Lhasa terrane
- Massif Central
- Moldanubian Plate (Austria)
- Moravo Silesian Plate (Czech Republic)
- Midland Craton (Great Britain)
- North Atlantic Craton
- North China Craton (China and Korea)
- Ossa-Morena Plate (Spain)
- Piemont-Liguria Plate
- Proto-Alps Terrane (Austria)
- Rhenohercynian Plate
- Sarmatian Craton (Ukraine)
- Sarmatian Craton
- Saxothuringian Plate
- Siberian Craton (Russia)
- South Portuguese Plate (Portugal)
- Tarim Craton (China)
- Teplá-Barrandian Terrane (Czech Republic)
- Ukrainian Shield (Ukraine)
- Valais Plate
- Volgo-Uralian Craton, (Russia)
- Yakutai Craton, eastern (Russia)
- Yangtze Craton (China)

Indo-Australian plate

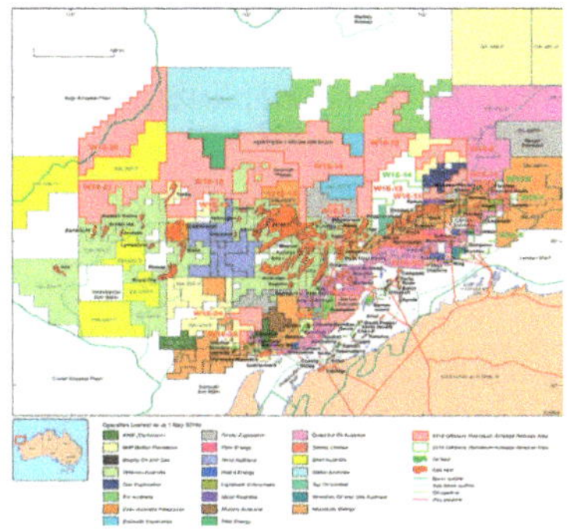

Basic geological regions of Australia, by age.

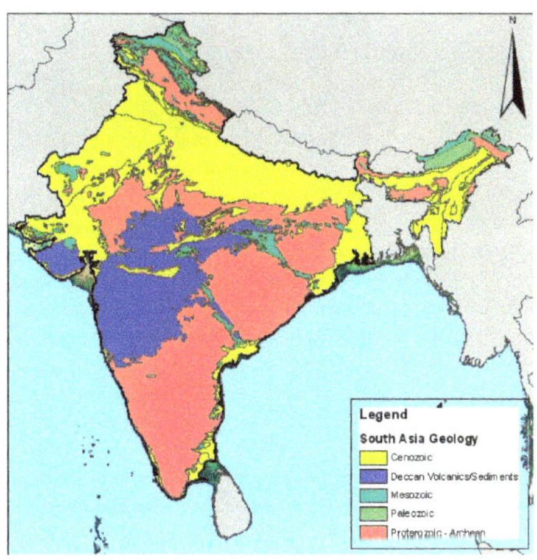

Map of chronostratigraphic divisions of India.

- Altjawarra Craton (Australia)
- Bhandara Craton, (India)
- Bundelkhand Craton, (India)
- Dharwar Craton, (India)
- Central Craton (Australia)
- Curnamona Craton (Australia)
- Gawler Craton (Australia)
- Indian Craton (India)
- Narooma Terrane (Australia)
- Pilbara Craton (Australia)

- Singhbhum Craton (India)
- Yilgarn Craton (Australia)
- Western Australian Shield (Australia)
- Zealandia, see Moa Plate and Lord Howe Rise (New Zealand)

North American Plate

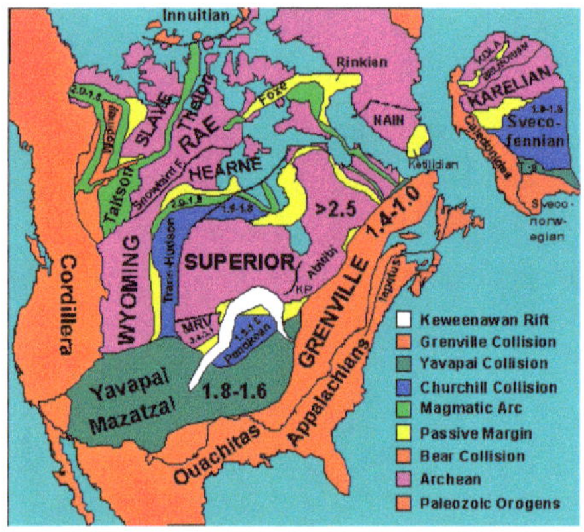

North American cratons and basement rocks.

- Avalonia (Canada, Great Britain, and United States)
- Carolina Plate
- Churchill Craton (Canada)
- Farallon Plate (split into the Cocos, Explorer, Juan de Fuca, Gorda Plates, Nazca Plate, and Rivera Plates)
- Florida Plate (United States)
- Hearne Craton (Canada)
- Laurentian Craton (Canada and United States)
- Insular Plate
- Intermontane Plate
- Izanagi Plate
- Mexican Plate (Mexico)
- Nain province (Canada)
- Newfoundland Plate (Canada)
- Nova Scotia Plate
- Rae Craton (Canada)
- Sask Craton (Canada)

- Sclavia Craton (Canada)
- Slave Craton (Canada)
- Superior Craton (Canada)
- Wyoming Craton (United States)

South American plate

- Amazonian craton (Brazil)
- Guiana Shield (Brazil, Colombia, French Guiana, Guyana, Suriname, and Venezuela)
- Rio Apas craton (Brazil, Paraguay)
- Río de la Plata craton (Argentina and Uruguay)
- São Francisco craton (Brazil)
- Arequipa–Antofalla craton (Argentina, Bolivia, Chile, Peru)

CIMMERIA

Tectonic History of Cimmeria

Cimmeria was an ancient continent, or, rather, a string of microcontinents or terranes, that rifted from Gondwana in the Southern Hemisphere and was accreted to Eurasia in the Northern Hemisphere. It consisted of parts of what is today Turkey, Iran, Afghanistan, Tibet, Shan–Thai, and Malaya. Cimmeria rifted from the Gondwanan shores of the Paleo-Tethys Ocean during the Carboniferous-earliest Permian and as the Neo-Tethys Ocean opened behind it, during the Permian, the Paleo-Tethys closed in front of it. Cimmeria rifted off Gondwana from east to west, from Australia to the eastern Mediterranean. It stretched across several latitudes and spanned a wide range of climatic zones.

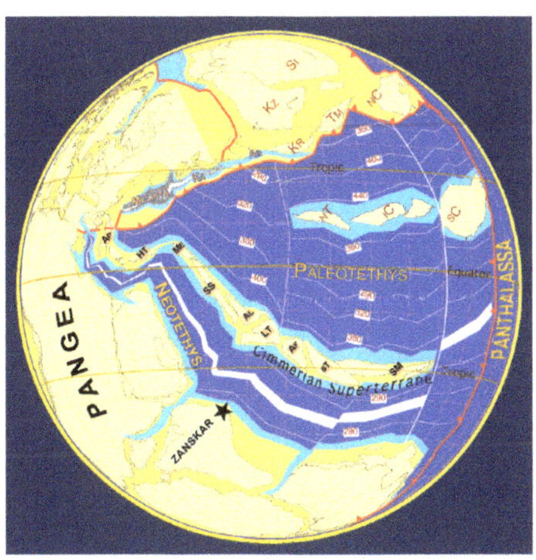

Cimmeria rifted off Gondwana's north-eastern shores around 250 Ma.

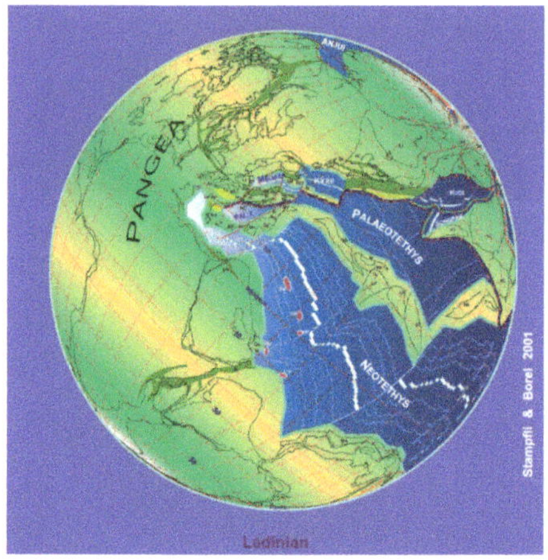

As Cimmeria migrated from Gondwana to Eurasia the Paleo-Tethys closed and the Neo-Tethys opened.

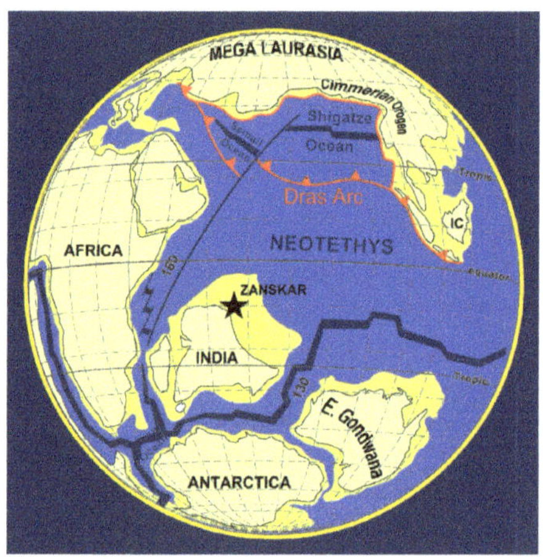

After 150 million years Cimmeria collided with Eurasia and the Cimmerian orogeny closed the Paleo-Tethys. As the break-up of Gondwana began in the south, the opening of the Indian Ocean initiated the closure of the Neo-Tethys.

HISTORY OF CONCEPT

First Concepts

A "large, ancient Mediterranean Sea" was first proposed by Austrian palaeontologist Melchior Neumayr in 1883. Studying the distribution of Jurassic faunas, he concluded that an equatorial ocean stretching from India to Central America must have separated a large continent in the northern hemisphere from one in the southern hemisphere. Austrian geologist Eduard Suess named this Mesozoic ocean the Tethys, a mythical ocean which separated a mythical continent — Gondwanaland, home of the tongue-shaped flora — from a boreal continent. German geophysicist Alfred Wegener, in contrast, developed a concept of a single, global continent – the supercontinent Pangea — which, in his view, didn't leave any room for an equatorial ocean. A wedge-shaped, east-facing Tethys within Pangea was, nevertheless, proposed by Australian geologist Samuel Warren Carey in 1958. This ocean was later identified as a succession of oceans separated by north-migrating terranes or continental blocks, one of which was Cimmeria.

Iranian Microcontinent

In 1974, after extensive field work in the Middle East, Swiss geologist Jovan Stöcklin identified the northern foot of the Alborz Range in northern Iran as the suture which in the Paleozoic was the northern shore of Gondwana and the remains of Paleo-Tethys Ocean. Stöcklin also noted that an early Mesozoic or late Paleozoic rift separated the Iranian Plate from the Arabian Plate, and that another southern suture must be the remains of the Neo-Tethys Ocean. The opening of this later ocean, Stöcklin realized, must have transformed Iran into a microcontinent. Those observations made Stöcklin the first to identify a small part of what would later be known as Cimmeria.

Stöcklin also noted that his proposal resembled the old concept of the world in which there were two continents, Angaraland in the north and Gondwana in the south, separated by an elongated ocean, the Tethys. Iran belonged to neither continent but was part of the realm of Tethys. Stöcklin›s southern suture was later confirmed by observations of the evolution of microflora in Iran; which had a Gondwanan affinity during the Carboniferous but a Eurasian affinity during the Late Triassic — Iran had clearly drifted from Gondwana to Laurasia.

Eurasian Superterrane

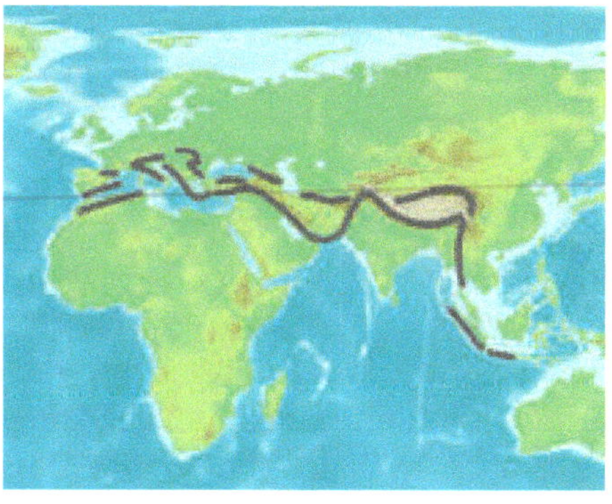

The Alpide belt is a system of sutures stretching across Eurasia within which the Cimmerian blocks are now located.

In the 1980s Turkish geologist Celâl Şengör finally extended Stöcklin›s Iranian microcontinent further west to Turkey and further east to Tibet and the Far East.Şengör also reused the name introduced by Suess in 1901, the Kimmerisches Gebirge – the Crimean or Cimmerian Mountains.

In the mountain range that now stretches from the Alps to Indonesia Şengör identified, using a simplified scheme, two distinct but superimposed orogenic systems containing a large number of anastomosing sutures: the older Cimmerides and the younger Alpides together forming what Şengör called the Tethysides super-orogenic system. These two orogenic systems are, thus, associated with two major periods of ocean closure: the earlier, northern, and much larger Cimmerides, and the later, southern, and smaller Alpides. Cimmeria was the long continental archipelago that separated the two oceans before the Paleo-Tethys closed.

This realm of Tethys thus covers most of Eurasia and a large time span (from north to south):

- Laurasia, Permian to Cretaceous
- Palaeo-Tethys, Early Carboniferous to Middle Jurassic
- Cimmeria, Triassic to Middle Jurassic
- Neo-Tethys, Permian or Triassic to Eocene, locally still extant
- Gondwana, Ordovician to Jurassic

This simple scheme, however, partly obscures the complex nature of the Tethyan cycles and terms such as Eocimmerian and Neocimmerian is often used for Late Triassic and Late Jurassic events respectively. Furthermore, a distinction is often made between two more recent Tethyan domains: the Alpine Tethys and the Neo-Tethys. The Alpine Tethys, the western domain in this scheme, separated south-western Europe from north-western Africa and was connected to the Central Atlantic. It is now completely closed and its suture encompasses the Maghrebids (stretching from Gibraltar to Sicily) as well as the Apennines and the Alps. The Neo-Tethys, the eastern domain, opened between Arabia and the Cimmerian terranes. The East Mediterranean Basin and the Gulf of Oman are considered relics of the Neo-Tethys which is thus still closing. These two domains were connected east of Sicily until the end of the Jurassic.

TECTONIC HISTORY

In the Late Paleozoic, when the Cimmerian blocks were still located on the northern margin of Gondwana, they were far away from any active margins and orogenic belts, but they had been affected by thermal subsidence since the Siluran opening of Paleo-Tethys. Carboniferous to Permian ophiolites along suture zones in Tibet and north-eastern Iran indicate that the active margin of Paleo-Tethys was located here. It was slab-pull forces in the Paleo-Tethys that detached Cimmeria from Gondwana and opened the Neo-Tethys. The mid-ocean ridge in the Paleo-Tethys subducted under Eurasia, as evidenced by Permian MORB (mid-ocean ridge basalt) in Iran. Slab roll-back in the Paleo-Tethys opened a series of back-arc basins along the Eurasian margin and resulted in the collapse of the Variscan cordillera. As the Paleo-Tethys subducted under the Eurasian southern margin, back-arc oceans formed from Austria to China. Some of these back-arcs closed during the Cimmerian orogeny (e.g. the Karakaya-Küre sequence of back-arc oceans in Turkey), others remained open (e.g. the Meliata-Maliac-Pindos back-arc oceans in the eastern Mediterranean) leading to the formation of younger back-arc oceans.

Turkey

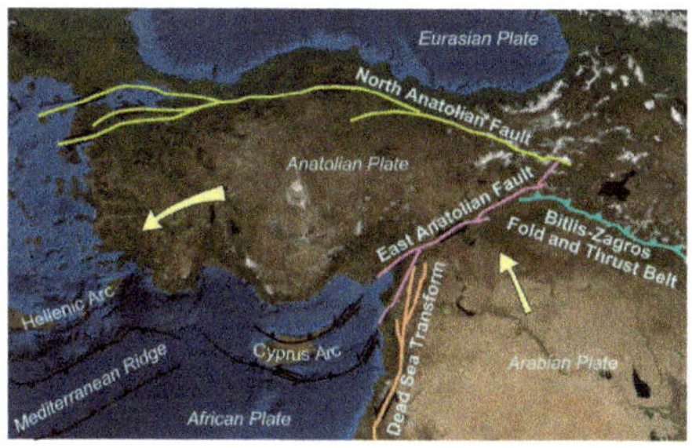

Geological map of Turkey.

Turkey is an assemblage of continental blocks that during the Permian were part of the northern margin of Gondwana. During the Permian-Triassic, as the Paleo-Tethys subducted under this margin (in what is today northern Turkey) a marginal sea opened and quickly got filled with sediments (today the basement of the Sakarya Composite Terrane in the Pontides). During the Late Triassic the Neo-Tethys began opening behind Cimmeria when the Eastern Mediterranean and its two eastern branches opened into the Bitlis-Zagros ocean (the southern branch of the Neo-Tethys).

During Early Jurassic Cimmeria began to disintegrate behind the Paleo-Tethyan volcanic arc. This opened the northern branch in the Neo-Tethys — the Intra-Pontide, Izmh-Ankara, and the Inner Tauride oceans. The closure of the Paleo-Tethys in the Middle Jurassic reduced the Cimmerian archipelago in Anatolia. South of the Cimmerian blocks there was now two branches of the Neo-Tethys, a northern, larger and more complex, and a southern, more reduced; the Anatolide-Tauride continent separated them, the small Sakarya continent was located within the northern branch. The Apulian continent was connected to the Anatolide-Tauride continent.

These Neo-Tethyan branches reached their maximum width during the Early Cretaceous, after which subduction under Eurasia gradually consumed them. During the Middle-Late Cretaceous this subduction opened a back-arc basin, the Western Black Sea Basin, which stretched west into the Balkans north of the Rhodope-Pontide island arc there. In the Cretaceous, this basin pushed the Istanbul terrane (near today's Istanbul) southward in front of it, from the Odessa Shelf in the north-western Black Sea. In the Eocene, the terrane finally collided with Cimmeria thereby ending the extension in the western Black Sea. Contemporaneously, the East Black Sea Basin opened when the East Black Sea Block was rotated counter-clockwise towards Caucasus.

In the late Jurassic collision between the Arabian Plate and the Bitlis-Pötürge fragment (today forming the Bitlis-Pötürge Massif in south-eastern Turkey) resulted in ophiolite obduction in both northern Arabia and northern Iran. After this collision the Tauride Ocean (north of the Bitlis-Pötürge fragment) closed and two basins opened north of subduction zone. The Rhodope-Pontide fragment and the Sakarya continent had amalgamated during Late Cretaceous and in the early-Late Eocene they finally collided with the Anatolide-Tauride continent — resulting in the Anatolides. The Arabian-Eurasian collision in eastern Turkey during the Late Eocene closed the two basins.

During the Paleogene Neo-Tethyan oceanic crust attached to the African Plate subducted along the Crete and Cyprus trenches. The Anatolide-Tauride continent collided with the Pontide and Kırşehir blocks in the Late Paleocene-Early Eocene. This closed the Ankara-Erzincan branches of the northern Neo-Tethys. During this closure, slab roll-back and break-off in the Eocene resulted in inversion in the Pontides and widespread magmatism in northern Turkey. Extension and upwelling followed, resulting in melting of lithospheric material beneath the Pontides.

In southern Turkey the northward subduction of the Neo-tethys along the Bitlis-Zagros subduction zone resulted in magmatism in the Maden-Helete arc (south-eastern Turkey) during the Late Cretaceous-Eocene and back-arc magmatism in the Taurides. The Bitlis-Zagros subduction zone finally closed in the Miocene and throughout the Oligocene-Neogene and Quaternary volcanism became increasingly localised. In the Late Oligocene, slab roll-back in the Hellenic Trench resulted in extension in the Aegean and western Turkey.

Iran

The subduction of western Neo-Tethys under Eurasia resulted in extensive magmatism in what is now northern Iran. In the Early Jurassic this magmatism had produced a slab pull force which contributed to the break-up of Pangea and the initial opening of the Atlantic. During the Late Jurassic-Early Cretaceous the subduction of the Neo-Tethys mid-ocean ridge contributed to the break-up of Gondwana, including the detachment of the Argo-Burma terrane from Australia. The Central-East Iranian Microcontinent (CEIM) sutured with Eurasia in the Late Triassic during the regional "Eocimmerian" orogenic event in northern Iran, but Iran is made of several continental blocks and the area must have seen a number of ocean closures in the Late Paleozoic and Early Mesozoic.

Caucasus

The Greater and Lesser Caucasus has a complicated geological history involving the accretion of a series terranes and microcontinents from the Late Precambrian to the Jurassic within the Tethyan framework. These include the Greater Caucasian, Black Sea-Central Transcaucasian, Baiburt-Sevanian, and Iran-Afghanistan terranes and island arcs. In the Caucasus region remnants of the Paleo-Tethys suture can be found in the Dzirula Massif which outcrops Early Jurassic sequences in central Georgia. It consists of Early Cambrian oceanic rocks and the possible remnants of a magmatic arc; their geometry suggests that suturing was followed by strike-slip faulting. Ophiolites also outcrop in the Khrami Massif in southern Georgia and another possible segment of the suture is present in the Svanetia region. The suture is older east of the Caucasus (northern Iran–

Turkmenistan) but younger both west of the Caucasus and further east in Afghanistan and the northern Pamirs.

Sibumasu

The easternmost part of Cimmeria, the Sibumasu terrane, remained attached to north-western Australia until 295–290 Ma when it began to drift northward, as supported by paleomagnetic and biogeographic data. The Qiangtang terrane was located west of Sibumasu and contiguous with it. Lower Permian layers in Sibumasu contain glacial-marine diamictites and Gondwanan faunas and floras which then developed independently before Sibumasu docked with Cathaysia. Sibumasu›s rapid northern journey is especially evident in the development of brachiopods and fusulinids.

The Baoshan terrane in western Yunnan, China, forms the northern part of Sibumasu. It is separated from the Burma Block by the Gaoligong Suture Zone to the west, and from the South China and Indochina continents in the east by the Chongshan Suture Zone and Changning-Menglian Belt. Like other parts of eastern Cimmeria, it was highly deformed by the intra-continental strike-slip faulting that followed the India-Asia collision.

Paleomagnetic data indicate South China and Indochina moved from near Equator to 20°N from the Early Permian to Late Triassic. Baoshan, in contrast, moved from 42°S in the Early Permian to 15°N in the Late Triassic. These blocks and terranes occupied similar paleo-latitudes during Late Triassic to Jurassic which indicates that they probably collided in the Late Triassic. This is also supported by geological evidence: 200-230 Ma granite in Lincang, near the Changning-Menglian suture, indicate a continent-continent collision occurred there in the Late Triassic; pelagic sediments in the Changning-Menglian-Inthanon ophiolite belt (between Sibumasu and Indochina) ranges in age from Middle Devonian to Middle Triassic, while, in the Inthanon suture, in contrast, Middle to Late Triassic rocks are non-pelagic with radiolaran cherts and turbidic clastics indicating the two blocks had at least approached each other by that time; volcanic sequences from the Lancangjiang igneous zone indicate a post-collisional setting had developed before the eruptions there around 210 Ma; and, the Sibumasu fauna developed from a non-marine peri-Gondwanan assemblage in the early Permian, to an endemic Sibumasu fauna in the Middle Permian, and to an Equatorial-Cathaysian in the Late Permian.

During the Early and Middle Palaeozoic Cimmeria was located at an Andean-style active margin. Glacial deposits and paleomagnetic data indicate that Qiangtang and Shan Thai-Malaya were still located far south adjacent to Gondwana during the Carboniferous. The equatorial fauna and flora of China indicate that it was separated from Gondwana during the Carboniferous.

Lhasa

The Lhasa terrane has been interpreted as part of Cimmeria and, if this is the case, must have rifted from Gondwana together with Sibumasu and Qiantang. The timing of Lhasa's northward drift is still controversial, however, and paleomagnetic data is extremely scarce. Sedimentological and stratigraphical evidence, for example, suggest that it separated from Gondwana in the Late Triassic when Qiantang was already being accreted to Eurasia. This proposed Late Triassic rifting of Lhasa has also been documented along the north-western shelf of Australia where the Western Burma and Woyla terranes eventually separated from Gondwana in the Late Jurassic.

Today the Bangong suture separates the Lhasa terrane from the Qiangtang terrane.

ECONOMIC IMPORTANCE

The present remains of Cimmeria, as a result of the massive uplifting of its continental crust, are unusually rich in a number of rare chalcophile elements. Apart from the Altiplano in Bolivia, almost all the world's deposits of antimony as stibnite are found in Cimmeria, with the major mines being in Turkey, Yunnan and Thailand. The major deposits of tin are also found in Malaysia and Thailand, whilst Turkey also has major deposits of chromite ore.

LHASA TERRANE

The **Lhasa terrane** is a terrane, or fragment of crustal material, sutured to the Eurasian Plate during the Cretaceous that forms present-day southern Tibet. It takes its name from the city of Lhasa in the Tibet Autonomous Region, China. The northern part may have originated in the East African Orogeny, while the southern part appears to have once been part of Australia. The two parts joined, were later attached to Asia, and then were impacted by the collision of the Indian Plate that formed the Himalayas.

Location

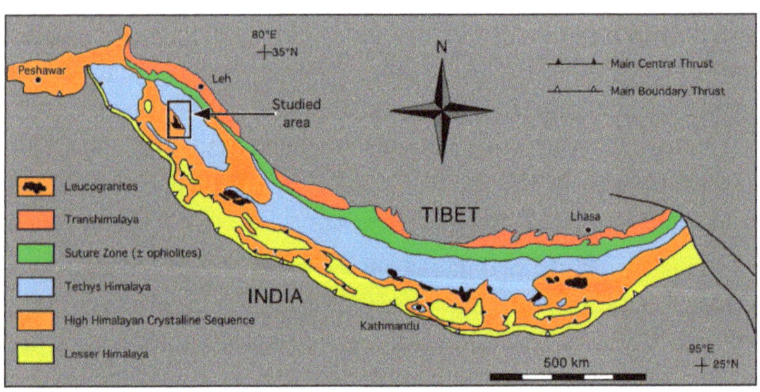

Tectonic units of the Himalaya. Green is the Indus-Yarlung suture zone. Red is the Transhimalaya. Lhasa to the east.

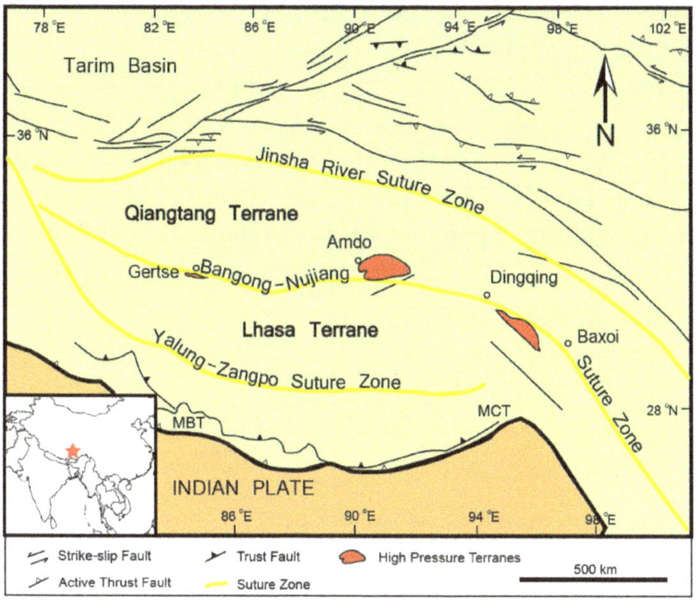

Lhasa and Qiangtang Terranes and associated suture zones.

The Lhasa terrane is separated from the Himalayas to the south by the Yarlung-Tsangpo suture, and from the Qiangtang terrane to the north by the Bangong-Nujiang suture. The Lhasa terrane has a Precambrian crystalline basement overlaid with sedimentary strata from the Paleozoic (c. 541–252 Ma[a]) and Mesozoic (c. 252–66 Ma) and containing magmatic rocks from the Paleozoic to Cenozoic (66 Ma to the present). It is thought to be the last crustal block to accrete to the Eurasian plate before it collided with the Indian plate in the Cenozoic.

Origins

The Lhasa terrane consisted of two blocks before the Mesozoic, the North Lhasa Block and the South Lhasa Block. The two blocks have lithology and detrital zircon ages similar to the Qiangtang terrane and to Tethyan strata in the Himalaya, which suggests these areas were nearby in Gondwana. The detrital zircon ages differ somewhat between the North and South Lhasa terranes. The South Lhasa terrane appears to have evolved as part of Australia in the late Precambrian and early Paleozoic. Isotopic analysis of detrital zircons of c. 1170 Ma from Paleozoic metasedimentary rocks in the Lhasa terrane shows identical values to detrital zircons of the same age from Western Australia. The detrital zircons probably came from southwest Australia's Albany-Fraser belt.

The North Lhasa terrane may have been formed in part from the northern part of the East African Orogeny. Neoproterozoic oceanic crustal rocks are included in the crystalline basement of the North Lhasa terrane, which are probably from the Mozambique Ocean that formed when the Rodinia super-continent broke up. In the Late Cryogenian, around 650 Ma, the oceanic crustal basement of North Lhasa experienced HP metamorphism in the subduction zone associated with the closing of the Mozambique ocean. In the

Early Paleozoic around 485 Ma it experienced MP metamorphism associated with the amalgamation of Eastern and Western Gondwana.

In the Early Paleozoic the North and South Lhasa terranes and the Qiangtang terrane experienced magmatism that seems to have been the result of an Andean-type orogeny caused when the Proto-Tethys Ocean was subducted after Gondwana was finally amalgamated. In the Middle Paleozoic around 360 Ma the Lhasa and Qiangtang terranes again experienced magmatism, apparently due to the subduction of the Paleo-Tethys Ocean.

Formation and Evolution

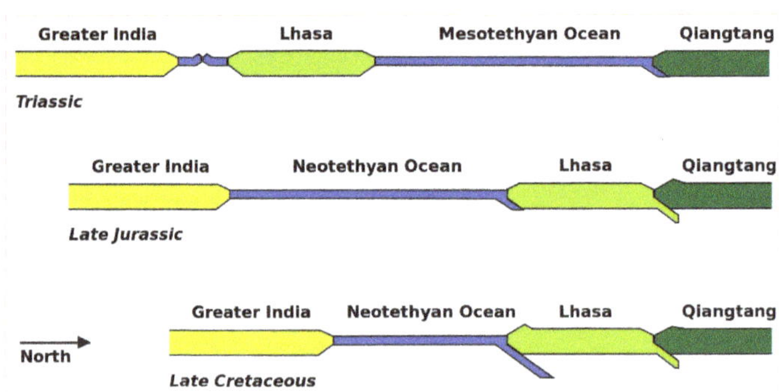

Lhasa terrane approach to Qiangtang terrane.

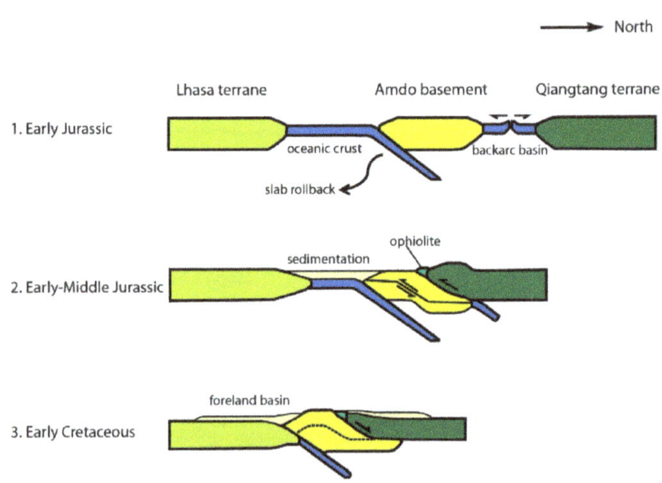

More detailed view of the tectonic evolution of the Bangong suture zone.

The Lhasa terrane was formed from the North and South Lhasa terranes, which were at first separated by the Paleo-Tethys Ocean, and were joined in a suture zone in the Late Paleozoic. The Paleo-Tethys Ocean that separated the North and South Lhasa terranes closed, and around 260 Ma in the Late Permian an HP metamorphic belt formed between the two blocks. Around 220 Ma in the Triassic an MP metamorphic belt formed.

The Tibetan Plateau was formed from a number of continental terranes that rifted from northern Gondwana in the Paleozoic and Mesozoic, moved northward and accreted to southern Asia. The Lhasa terrane is the farthest south of these terranes. The Lhasa terrane moved northward and collided with the Qiangtang terrane along the Banggongco-Nujiang Suture. The collision began towards the end of the late Jurassic (c. 163–145 Ma), and collision activity continued until the early Late Cretaceous (c. 100–66) Ma . During this period the terrane may have been shortened by at least 180 kilometres (110 mi). Strata from the Lower Jurassic in the Bangong suture between the Lhasa and Qiangtang terranes differ from the rocks in the Lhasa terrane and appear to have a unique source.

The collision with the Qiangtang terrane caused a peripheral foreland basin to form in the north part of the Lhasa terrane, which persisted into the Early Cretaceous. In some parts of the foreland basin the north-dipping subduction of the Neotethyan oceanic crust below the Lhasa terrane caused volcanism. The Gangdese volcanic arc was formed as this subduction continued along the southern margin of the Lhasa terrane. The Gangdese batholith intrudes the southern half of the Lhasa terrain. There is evidence that by the end of the Cretaceous the Southern Tibet crust was approximately twice as thick as normal.

Clastic sediments found in the terrane were deposited in shallow waters during the Early Cretaceous (c. 146–100 Ma.) In northern Lhasa these sediments formed in the foreland basin created during the Lhasa–Qiangtang collision. They are overlaid by marine limestone from the Aptian-Albian period, deposited in a shallow continental seaway. The Takena Formation developed in the Late Cretaceous in the foreland basin to the north of the Gangdese magmatic arc, and consists of marine limestone overlaid by fluvial red beds. Outcropped folds in the Takena Formation between Lhasa and Yangbajain are upright or lean slightly to the north or south, and indicate 30% to 50% shortening in the Late Cretaceous before the Indian collision.

India–Asia Collision

Contact with India began along the Yarlung-Zangbo suture around 50 Ma during the Eocene, and the two continents continue to converge. Magmatism continued in the Gangdese arc until as late as 40 Ma. There are competing hypotheses about the details of the tectonic processes during the collision between the Indian and Eurasian plates. At one extreme, some consider that during the collision the Indian crust was underthrust beneath the southern Asian crust, or injected into this crust. At the other extreme, some consider that the convergence was mostly accommodated by shortening of the Asian crust.

The results of seismic reflection profiling, reported in 1998, indicate that there may be a midcrustal partial-melt zone under the length of the Yangbajain-Damxung graben starting at a depth of 12 to 18 kilometres (7.5 to 11.2 mi). The reflection undulates, so the melt zone may have been tectonically deformed. North-dipping reflections deep in the crust below the Gangdese batholith at a depth of 40 to 60 kilometres (25 to 37 mi) may mark the downdip of the Yarlung-Zangbo suture, or may mark a more recent reverse fault. Taken together, the results indicate that the upper crust of the Lhasa terrane was moderately shortened by the collision, with melting in the middle crust. They neither support nor rule out underthrusting or fluid injection of the Indian continental crust below the Lhasa terrane.

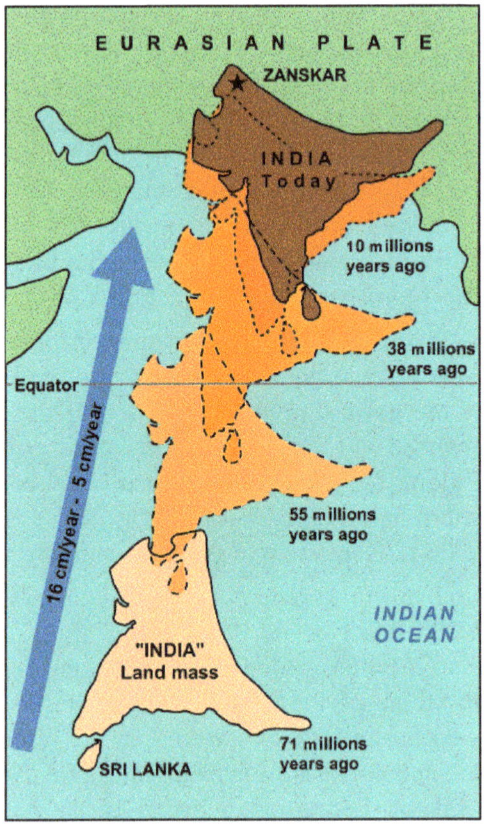

India-Eurasia collision 70-0 Ma.

The Linzizong Formation is distributed widely along the Gangdese Belt. It was emplaced between 69 and 43 Ma near Lhasa and between 54 and 37 Ma in southwestern Tibet. It is slightly folded and slopes gently to the north. The formation is unconformably underlain by Cretaceous sedimentary sequences more than 3,000 metres (9,800 ft) thick, which are strongly folded. The results of palaeomagnetic studies of the Linzizong Formation in the Linzhou Basin and the Takena Formation reported in 2009 indicate that there was little movement of the Lhasa terrane in the Cretaceous and Early Eocene. The measurements give a northward movement of the Lhasa terrane since then of 1,847 ± 763 kilometres (1,148 ± 474 mi). This implies that there was significant crustal shortening as the collision progressed. The South Lhasa terrane experienced metamorphism and magmatism in the Early Cenozoic (55–45 Ma) and metamorphism in the Late Eocene (40–30 Ma), presumably due to the collision between the continents of India and Eurasia.

Strata

Sedimentary strata from the Palaeozoic are mainly Carboniferous sandstone, metasandstone, shale and phyllite, and lesser Ordovician, Silurian and Permian limestone. Precambrian strata are rarely exposed. Rocks from the Triassic include inter-bedded limestone and basaltic volcanic units, most common along the terrane's southern margin. In the northern terrane the Jurassic strata are deepwater sandstone and shale, often with ophiolitic assemblages. In the southern terrane the Jurassic strata are marine limestone and mudstone. Strata from the Lower Cretaceous are clastic mudstone, sandstone and local conglomerate units. The Lower Cretaceous clastic units are overlaid by a shallow marine limestone from the Aptian-Albian period, exposed in many places, which in some places holds Cenomanian fossils. The strata from the Upper Cretaceous are successions of arkosic fluvial sandstone and mudstone.

AGULHAS BASIN

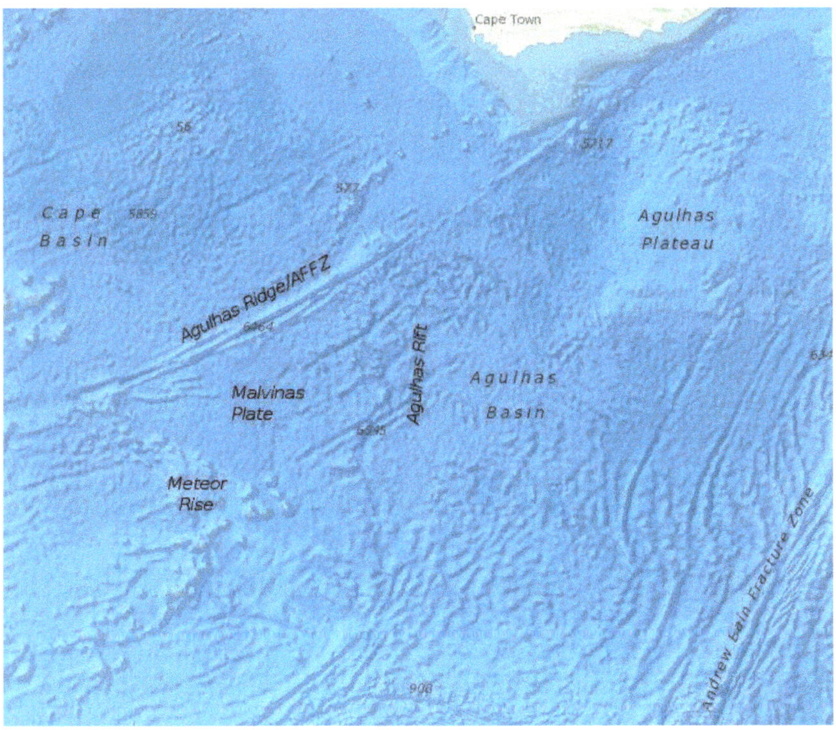

The Agulhas Basin and some of the bathymetric structures mentioned in the text.

The **Agulhas Basin** is an oceanic basin located south of South Africa where the South Atlantic Ocean and south-western Indian Ocean meet. Part of the African Plate, it is bounded by the Agulhas Ridge (part of the Agulhas-Falkland Fracture Zone) to the north and the Southwest Indian Ridge to the south; by the Meteor Rise to the west and the Agulhas Plateau to the east. A large number of bathymetric anomalies hints at the basin's dynamic tectonic history.

In a Late Paleocene (59-56 Ma) reconstruction of the opening of the South Atlantic Ocean (i.e. the separation of South America and Africa during the Gondwana break-up) the Meteor Rise lies conjugate to the Islas Orcadas Rise (east of the Falkland Plateau). The separation of the Meteor Rise and the Islas Orcadas Rise marks the beginning of the formation of the Agulhas Basin.

The Agulhas Ridge extends from the northern tip of the Meteor Rise towards the Agulhas Bank south of South Africa. The ridge, however, ends abruptly in a small plateau at 40°S 15°E where it intersects a northeastward-trending spreading centre (the Agulhas Rift) that was abandoned during the Early Paleocene (61 Ma). The presence of a short-lived tectonic plate between these structures was first proposed by LaBrecque & Hayes 1979. They named it the Malvinas Plate and proposed that it was active from 90 Ma until the spreading ceased in the Agulhas Basin at 65 Ma. The plate is located at a proto-Bouvet Triple Junction.

Marks & Stock 2001 found that Late Cretaceous (100-66 Ma) fracture zones generated on the Agulhas Rift do not align with those north of the Agulhas Fracture Zone and cannot therefore have been formed by the spreading of South America and Africa. Furthermore, magnetic anomalies on the Malvinas Plate do not align with their conjugates on the African Plate if the spreading rates and directions of South America and Africa are used as a guide. They also noted that the Agulhas Fracture Zone do not lay perpendicular to traces of the South America-Africa spreading north of it and cannot, therefore, have been generate by this spreading.

The Agulhas Rift is the abandoned Malvinas-Africa ridge crest. 97 Ma the plate boundary in the Agulhas Basin was reorganised when the Mid-Atlantic Ridge made an eastward jump. This brought the boundary towards the Agulhas Plateau where excessive volcanism was building a large igneous province. The inception of the Malvinas Plate accompanied this shortening of the Agulas Fracture Zone. 61 Ma, the Malvinas Plate was finally incorporated into the African Plate when the Malvinas-Africa ridge was abandoned as the result of a westward ridge jump along the Agulhas-Falkland Fracture Zone. This second ridge jump reduced one of the most spectacular fracture zones in Earth›s history — 1,200 km (750 mi) in length — to 180 km (110 mi).

Oceanography

The Agulhas Current flows south along the African east coast. When it reaches the southern tip of Africa, it retroflects back into the Indian Ocean. At this retroflection it leakes warm core eddies known as Agulhas rings into the South Atlantic. This mesoscale anti-cyclonic rings feed the Atlantic meridional overturning circulation (AMOC) and are therefore believed to affect the global climate, although the connection between the leakage, AMOC, and climate change is still poorly understood. In the Agulhas basin, half of these rings are subdivided one or several times. A majority of the subdivided rings split at or near the bathymetric obstacles on the western side of the Agulhas Basin but almost a quarter of them also merge there either because of the obstacles or because of intense interaction with other rings.

Antarctic Bottom Water (AABW) originated in the Oligocene with the opening of the Drake Passage and the Tasmanian Seaway and resulted in the thermal insulation of Antarctica. AABW mixes with other masses in the Antarctic Circumpolar Current (ACC) to form the Circumpolar Deep Water (CDW). In the Agulhas Basin CDW flows northward and is deflect mostly westward by the Agulhas Ridge. A branch of CDW, however, enters the Cape Basin west of the ridge from where it flows west along the northern side of the ridge before being deflected north-east (along the Walvis Ridge) at the eastern end of the ridge. This long detour in the Cape Basin and mixing with North Atlantic Deep Water (NADW) results in warmer water than other CDW masses. Antarctic Intermediate Water (AAIW) flows above the CDW in the Agulhas Basin in an anti-cyclonic path (in contrast to the cyclonic path followed by CDW.)

FARALLON PLATE

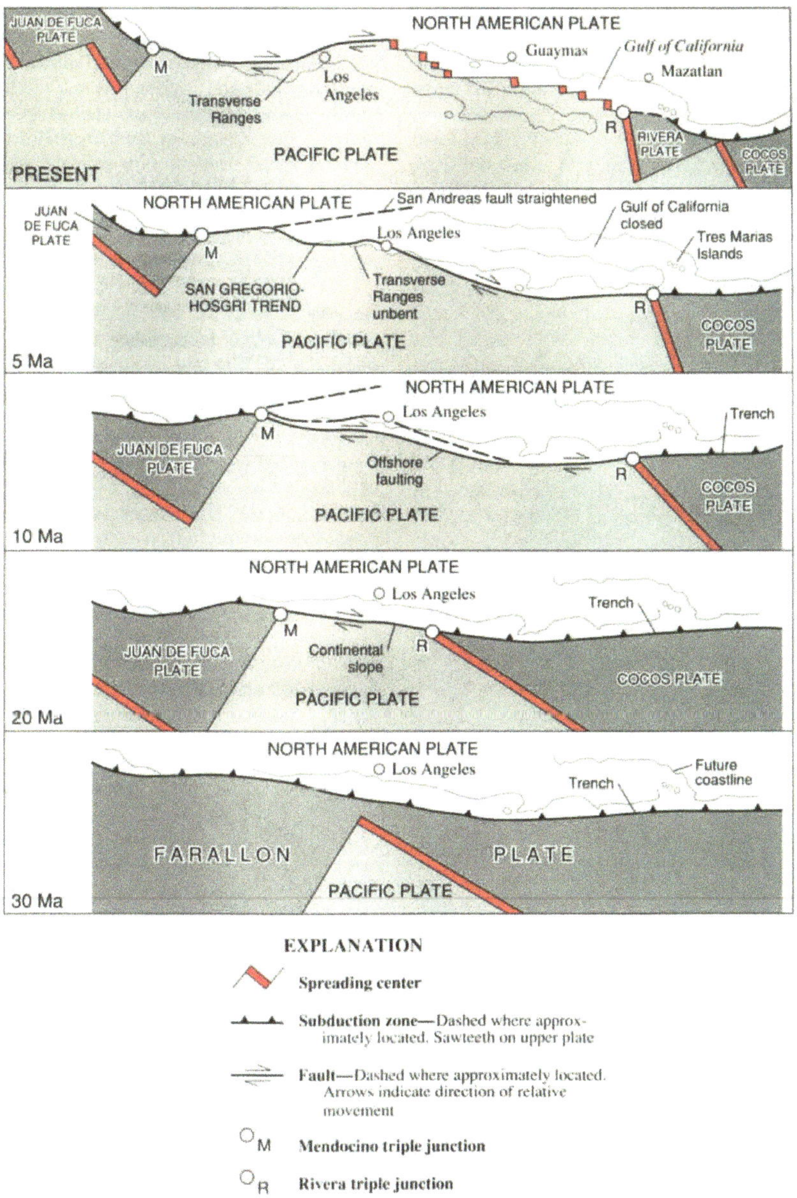

Formation of the Juan de Fuca (including Explorer and Gorda) and Cocos plates (including Rivera) and of the San Andreas Fault from the Farallon plate.

The **Farallon Plate** was an ancient oceanic plate that began subducting under the west coast of the North American Plate—then located in modern Utah—as Pangaea broke apart during the Jurassic period. It is named for the Farallon Islands, which are located just west of San Francisco, California.

Over time, the central part of the Farallon Plate was completely subducted under the southwestern part of the North American Plate. The remains of the Farallon Plate are the Juan de Fuca, Explorer and Gorda Plates, subducting under the northern part of the North American Plate; the Cocos Plate subducting under Central America; and the Nazca Plate subducting under the South American Plate.

The Farallon Plate is also responsible for transporting old island arcs and various fragments of continental crustal material rifted off from other distant plates and accreting them to the North American Plate.

These fragments from elsewhere are called terranes (sometimes, exotic terranes). Much of western North America is composed of these accreted terranes.

Current State

The understanding of the Farallon Plate is rapidly evolving as details from seismic tomography provide improved details of the submerged remnants. Since the North American west coast shows a convoluted structure, significant work has been required to resolve the complexity. In 2013 a new and more nuanced explanation emerged, proposing two additional now-subducted plates which would account for some of the complexity.

As data accumulated, a common view developed that one large oceanic plate, the Farallon plate, acted as a conveyor belt, conveying terranes to North America's west coast, where they accreted. As the continent overran the subducting Farallon plate, the denser plate became subducted into the mantle below the continent. When the plates converged, the dense oceanic plate sank into the mantle to form a slab below the lighter continent.

Farallon Plate subduction forms Cordilleran North America

As of 2013, it is generally accepted that the western quarter of North America consists of accreted terrane accumulated over the past 200 million years as a result of the oceanic Farallon plate moving terranes onto the continental margin as it subducts under the continent. However this simple model was unable to explain many terrane complexities, and is inconsistent with seismic tomographic images of subducting slabs penetrating the lower-mantle. In April 2013 Sigloch and Mihalynuk noted that under North America these subducting slabs formed massive, essentially vertical walls of 800 km to 2,000 km deep and 400–600 km wide, forming slab walls. One such large slab wall runs from north-west Canada to the eastern U.S. and extends to Central America; this slab wall had traditionally been associated with the subducting Farallon plate. Sigloch and Mihalynuk proposed that the Farallon should be partitioned into Northern Farallon, Angayucham, Mezcalera and Southern Farallon segments based on recent tomographic models. Under this model, the North American continent overrides a series of subduction trenches and incorporates microcontinents (similar to those in the modern-day Indonesian archipelago) as it moves west in the following sequence:

- 165–155 Myr ago the Mezcalera promontory (the leading terrane to strike North America) strikes land and begins to be overridden. The overridden segment is replaced by an incipient South Farallon trench.
- 160–155 Myr ago the Rocky Mountain deformation begins, recorded by a synorogenic (formed contemporaneously with the orogen) clastic wedge. The Franciscan subduction complex on the South Farallon plate begins.
- 125 Myr ago the collision of the North America margin with an archipelago of terranes (Mezcalera / Angayucham /Southern Farallon island arcs) begins. This broad expanse causes strong deformations and creates the Sevier Mountains and the Canadian Rocky Mountains.
- 124–90 Myr ago the Omenica magmatic belts are formed in the Pacific Northwest along with a gradual override of the Mezcalera promontory by the Pacific Northwest.
- 85 Myr ago the South Farallon trench moves westward after accretion of the Shatsky Rise Conjugate plateau. Sonora volcanism results from the slab sinking. The Tarahumara ignimbrite province is formed.
- 85–55 Myr ago Strong transpressive coupling of Farallon plate to terranes produces the buoyant Shatsky Rise. The Laramide orogeny results from basement uplift more than 1,000 km inland.
- 72–69 Myr ago the Angayucham arc, is overridden by North America and Carmacks volcanic episode results.
- 85–55 Myr ago Conjugate subducts. Northward shuffle of Insular terrane, Intermontane terrane, and Angayucham terranes along margin.
- 55–50 Myr ago saw the override of the Cascadia Root arc by the Pacific Northwest along with accretion of the Siletzia and Pacific Rim terranes.
- 55–50 Myr ago Final override of westernmost Angayucham occurred, with an explosive end of Coast Mountain arc volcanism

When the final archipelago, the Siletzia archipelago lodged as a terrane, the associated trench stepped west as the terrane accreted, converting an intra-oceanic subduction trench into the current Cascadia subduction zone and creating a slab window.

REFERENCES

1. Reconstruction from Dèzes 1999, p. 16
2. Reconstruction from Stampfli & Borel 2002, p. 27
3. Scotese & McKerrow 1990, pp. 4, 5, 17
4. Golonka 2007, p. 182
5. Stampfli & Borel 2002, pp. 24, 28
6. Metcalfe 2002, p. 556
7. Neumayr 1883
8. Suess 1893; Suess 1901

9. Hsü & Bernoulli 1978, Paleotethys, pp. 943–944 and references therein including Carey 1958
10. Stöcklin 1974, Introduction, p. 873
11. Stampfli 2000, Some definitions, pp. 1–2
12. Şengör 1984, Şengör 1987
13. Suess 1901, p. 22
14. Şengör et al. 1988, pp. 119–120, 123
15. See for example Frizon de Lamotte et al. 2011, The Zagros Domain and Its Arabian Foreland, p. 4
16. Frizon de Lamotte et al. 2011, Introduction, pp. 1, 4
17. Stampfli et al. 2001, Initial Conditions, pp. 57–58
18. Şengör & Yilmaz 1981, Abstract
19. Okay, Şengör & Görür 1994, Abstract; Fig. 3, p. 269
20. Richards 2015, Turkey, pp. 329-330
21. Buchs et al. 2013, Introduction, pp. 267–268
22. Gamkrelidze & Shengelia 2007, Introduction, p. 57
23. Şengör et al. 1988, pp. 139–140
24. Metcalfe 2002, p. 556; Position of the Sibumasu Terrane, pp. 562–563; Position of the Qiangtang Terrane, p. 563
25. Zhao et al. 2015, Geological setting and sampling, p. 3
26. Zhao et al. 2015, The closure of the East Paleotethys Ocean, pp. 10–11, 13
27. Metcalfe 2002, Position of the Lhasa Terrane, p. 563
28. Metcalfe 1996, Late Triassic to Late Jurassic rifting, pp. 104–105
29. Ma – Millions of years ago
30. Ozacar 2015.
31. Zhang et al. 2014, p. 170–171.
32. Wan 2010, p. 139.
33. Leier et al. 2007, p. 361.
34. Di et al. 2011.
35. Wan 2010, p. 210.
36. Metcalfe 1994, pp. 97–111.
37. Leier 2005.
38. Leier et al. 2007, p. 363.
39. Alsdorf, BrownNelson & Makovsky 1998, p. 502.
40. Leier et al. 2007, p. 363–364.
41. Alsdorf, BrownNelson & Makovsky 1998, p. 501.
42. Liebke et al. 2010, p. 1200.
43. Liebke et al. 2010, p. 1199.

Geology of Venus

Chapter Contents

Venus and Earth are near-twins in size, mass, composition, and conditions when they formed. Like Earth, Venus has a large iron core and rocky silicate mantle. Its crust is largely basalt (again, like Earth). Now we have looked in detail at the earth, we compare the overall behavior of Venus. Much of our knowledge is based on the exquisitely detailed surface mapping by radar spacecraft. These maps show that the surface of this planet has not developed plates, and as a result the interior heat has dramatically different ways of rising to the surface compared with the earth.

INTRODUCTION

Venus is a planet with striking geology. Of all the other planets in the Solar System, it is the one nearest to Earth and most like it in terms of mass, but has no magnetic field or recognizable plate tectonic system. Much of the ground surface is exposed volcanic bedrock, some with thin and patchy layers of soil covering, in marked contrast with Earth, the Moon, and Mars. Some impact craters are present, but Venus is similar to Earth in that there are much fewer craters than on the other rocky planets that are largely covered by them. This is due in part to the thickness of the Venusian atmosphere disrupting small impactors before they strike the ground, but the paucity of large craters may be due to volcanic re-surfacing, possibly of a catastrophic nature. Volcanism appears to be the dominant agent of geological change on Venus. Some of the volcanic landforms appear to be unique to the planet. There are shield and composite volcanoes similar to those found on Earth. Given that Venus has approximately the same size, density, and composition as Earth, it is plausible that volcanism may be continuing on the planet today, although convincing direct observations of a volcanic eruption have not yet occurred, leaving modern volcanism an open question.

Radar global map of the surface of Venus.

Most of the Venusian surface is relatively flat; it is divided into three topographic units: lowlands, highlands, and plains. In the early days of radar observation the highlands drew comparison to the continents of Earth, modern research has shown that this is superficial and the absence of plate tectonics makes this comparison misleading. Tectonic features are present to a limited extent, including linear deformation belts composed of folds and faults. These may be caused by mantle convection. Many of the tectonic features such as *tesserae* (large regions of highly deformed terrain, folded and fractured in two or three dimensions), and *arachnoids* (for those features resembling a spider's web) are associated with volcanism.

Eolian landforms are not widespread on the planet's surface, but there is considerable evidence the planet's atmosphere causes the chemical weathering of rock, especially at high elevations. There is only 20 ppm water vapor in the Venusian atmosphere, and no liquid erosion features (aside from those of lava flows) are visible in radar images of the surface. The atmosphere shows isotopic evidence of having been stripped of volatile elements by offgassing and solar wind erosion over time, implying the possibility that Venus may have had liquid water at some point in the distant past; no direct evidence for this has been found. Much speculation about the geological history of Venus continues today.

The surface of Venus is not easily accessible because of the extremely thick atmosphere (some 90 times that of Earth›s) and the 470 °C (878 °F) surface temperature. Much of what is known about it stems from orbital radar observations, because the surface is permanently obscured in visible wavelengths by cloud cover. In addition, a number of landers have returned data from the surface, including images.

TOPOGRAPHY

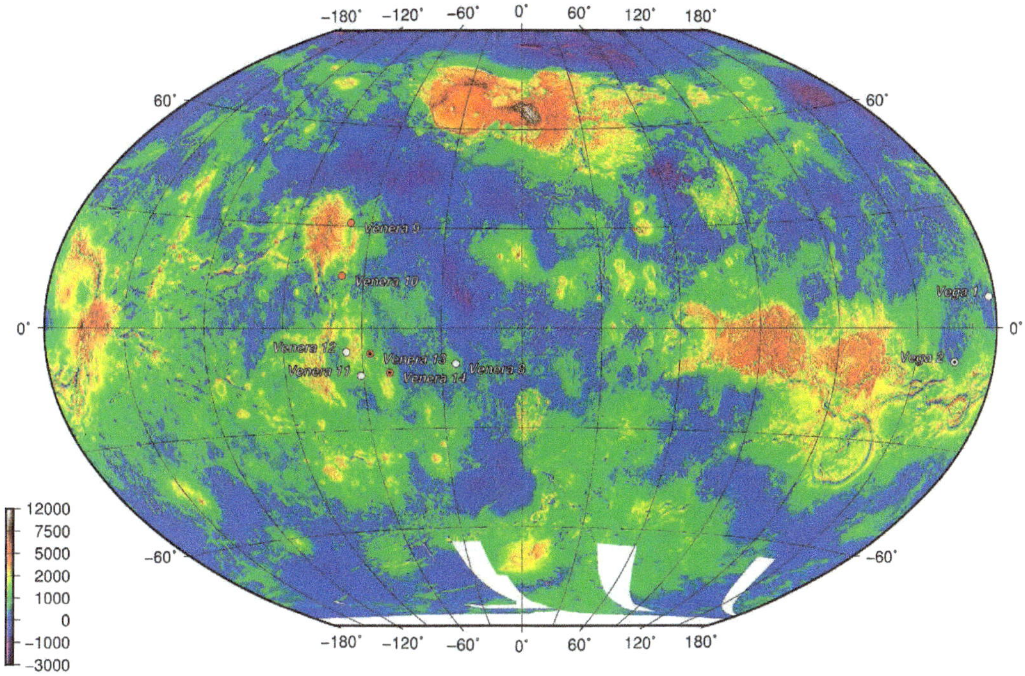

Venus topography.

The surface of Venus is comparatively flat. When 93% of the topography was mapped by *Pioneer Venus*, scientists found that the total distance from the lowest point to the highest point on the entire surface was about 13 kilometres (8.1 mi), about the same as the vertical distance between the Earth's ocean floor and the higher summits of the Himalayas, about 13 kilometres (8.1 mi). This similarity is to be expected as the maximum attainable elevation contrasts on a planet are largely dictated by the strength of the planet›s gravity and the mechanical strength of its lithosphere, these are similar for Earth and Venus.

According to data from the *Pioneer* altimeters, nearly 51% of the surface is located within 500 meters (1640 feet) of the median radius of 6,052 km (3,761 mi); only 2% of the surface is located at elevations greater than 2 kilometres (1.2 mi) from the median radius.

The altimetry experiment of *Magellan* confirmed the general character of the landscape. According to the *Magellan* data, 80% of the topography is within 1-kilometre (0.62 mi) of the median radius. The most important elevations are in the mountain chains that surround Lakshmi Planum: Maxwell Montes (11 km, 6.8 mi), Akna Montes (7 km, 4.3

mi) and Freya Montes (7 km, 4.3 mi). Despite the relatively flat landscape of Venus, the altimetry data also found large inclined plains. Such is the case on the southwest side of Maxwell Montes, which in some parts seems to be inclined some 45°. Inclinations of 30° were registered in Danu Montes and Themis Regio.

About 75% of the surface is composed of bare rock.

Based on altimeter data from the *Pioneer Venus* probe, supported by 'Magellan' data, the topography of the planet is divided into three provinces: lowlands, deposition plains, and highlands.

Highlands

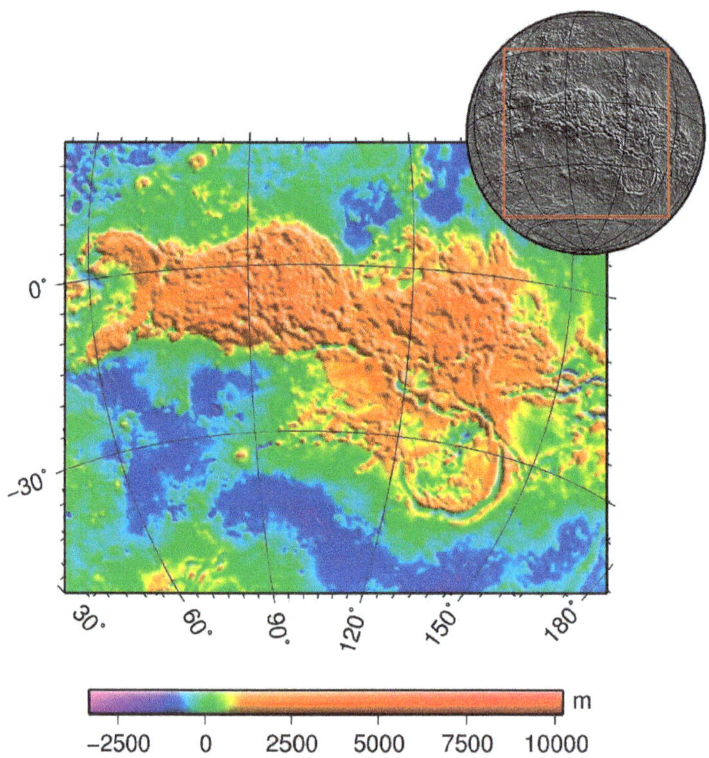

Aphrodite Terra topography.

This unit covers about 10% of the planet's surface, with elevations greater than 2 km. The most important provinces of the highlands are Aphrodite Terra, Ishtar Terra, and Lada Terra, as well as the regions Beta Regio, Phoebe Regio and Themis Regio. The regions Alpha Regio, Bell Regio, Eistla Regio and Tholus Regio form a less important group of highlands.

Some of the terrain in these areas is unusually efficient at reflecting radar signals. This is possibly analogous to snow lines on Earth and is likely related to temperatures and pressures there being lower than in the other provinces due to the higher elevation, which allows for distinct mineralogy to occur. It is thought that high-elevation rock formations may contain or be coated by minerals that have high dielectric constants. The high dielectric minerals would be stable at the ambient temperatures in the highlands, but not on the plains that comprise the rest of the planet's surface. Pyrite, an iron sulfide, matches these criteria and is widely suspected as a possible cause; it would be produced by chemical weathering of the volcanic highlands after long-term exposure to the sulfur-bearing Venusian atmosphere. The presence of pyrite on Venus has been contested, with atmospheric modeling showing that it might not be stable under Venusian atmospheric conditions. Other hypotheses have been put forward to explain the higher radar reflectivity in the highlands, including the presence of a ferroelectric material whose dielectric constant changes with temperature (with Venus having a changing temperature gradient with elevation). It has been observed that the character of the radar-bright highlands is not consistent across the surface of Venus. For example, Maxwell Montes shows the sharp, snow line-like change in reflectivity that is consistent with a change in mineralogy, whereas Ovda Regio shows a more gradual brightening upwards trend. The brightening upwards trend on Ovda Regio is consistent with a ferroelectric signature, and has been suggested to indicate the presence of chlorapatite.

Deposition plains

Deposition plains have elevations averaging 0 to 2 km and cover more than half of the planet's surface.

Lowlands

The rest of the surface is *lowlands* and generally lies below zero elevation. Radar reflectivity data suggest that at a centimeter scale these areas are smooth, as a result of *gradation* (accumulation of fine material eroded from the highlands).

SURFACE OBSERVATIONS

Ten spacecraft have successfully landed on Venus and returned data, all were flown by the Soviet Union. Venera 9, 10, 13, and 14 had cameras and returned images of soil and rock. Spectrophotometry results showed that these four missions kicked up dust clouds on landing, which means that some of the dust particles must be smaller than about 0.02 mm. The rocks at all four sites showed fine layers, some layers were more reflective than others. Experiments on rocks at the Venera 13 and 14 sites found that they were porous and easily crushed (bearing maximum loads of 0.3 to 1 Mpa. these rocks may be weakly lithified sediments or volcanic tuff. Spectrometry found that the surface materials at the Venera 9, 10, 14 and Vega 1 and 2 landing had chemical compositions similar to tholeiitic basalts, while the Venera 8 and 13 sites chemically resembled alkaline basalts.

IMPACT CRATERS AND AGE ESTIMATES OF THE SURFACE

Colorized radar image of Danilova crater in relief.

Earth-based radar surveys made it possible to identify some topographic patterns related to craters, and the *Venera 15* and *Venera 16* probes identified almost 150 such features of probable impact origin. Global coverage from *Magellan* subsequently made it possible to identify nearly 900 impact craters.

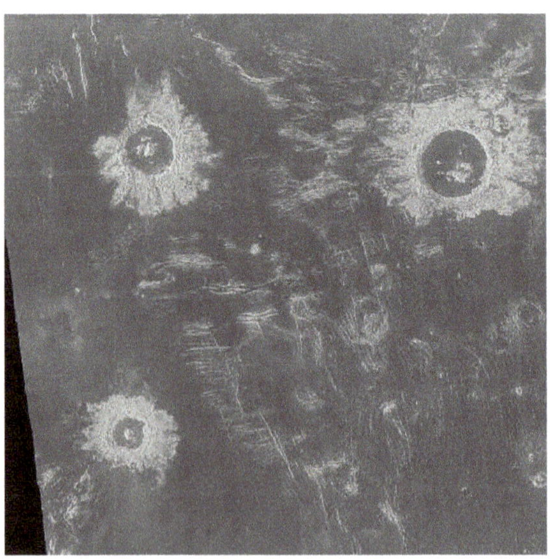

Danilova, Aglaonice and Saskja craters.

Compared to Mercury, the Moon and other such bodies, Venus has very few craters. In part, this is because Venus's dense atmosphere burns up smaller meteorites before they hit the surface. The *Venera* and *Magellan* data are in agreement: there are very few impact craters with a diameter less than 30 kilometres (19 mi), and data from *Magellan* show an absence of any craters less than 2 kilometres (1.2 mi) in diameter. The small craters are irregular and appear in groups, thus pointing to the deceleration and the breakup of impactors. However, there are also fewer of the large craters, and those appear relatively young; they are rarely filled with lava, showing that they were formed after volcanic activity in the area ceased, and radar data indicates that they are rough and have not had time to be eroded down.

Compared to the situation on bodies such as the Moon, it is more difficult to determine the ages of different areas of the surface on Venus, on the basis of crater counts, due to the small number of craters at hand. However, the surface characteristics are consistent with a completely random distribution, implying that the surface of the entire planet is roughly the same age, or at least that very large areas are not very different in age from the average.

Taken together, this evidence suggests that the surface of Venus is geologically young. The impact crater distribution appears to be most consistent with models that call for a near-complete resurfacing of the planet. Subsequent to this period of extreme activity, process rates declined and impact craters began to accumulate, with only minor modification and resurfacing since.

A young surface all created at the same time is a different situation compared with any of the other terrestrial planets.

GLOBAL RESURFACING EVENT

Age estimates based on crater counts indicate a young surface, in contrast to the much older surfaces of Mars, Mercury, and the Moon. For this to be the case on a planet without crustal recycling by plate tectonics requires explanation. One hypothesis is that Venus underwent some sort of global resurfacing about 300–500 million years ago that erased the evidence of older craters.

One possible explanation for this event is that it is part of a cyclic process on Venus. On Earth, plate tectonics allows heat to escape from the mantle by advection, the transport of mantle material to the surface and the return of old crust to the mantle. But Venus has no evidence of plate tectonics, so this theory states that the interior of the planet heats up (due to the decay of radioactive elements) until material in the mantle is hot enough to force its way to the surface. The subsequent resurfacing event covers most or all of the planet with lava, until the mantle is cool enough for the process to start over.

VOLCANOES

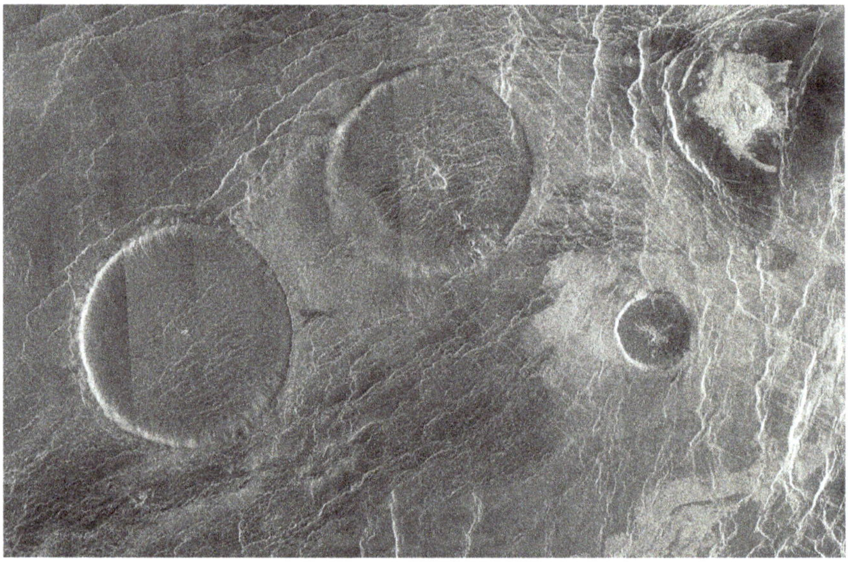

Radar image of pancake domes in Venus's Eistla region. The two larger ones are approximately 65 km (40 mi) wide and rise less than 1 km (0.62 mi) above the surrounding plain. These wide and fairly low, flat-topped volcanoes are a type of landform that is unique to Venus. They were probably formed by extrusions of highly viscous lava that was too sticky to flow very far down-slope from their vents.

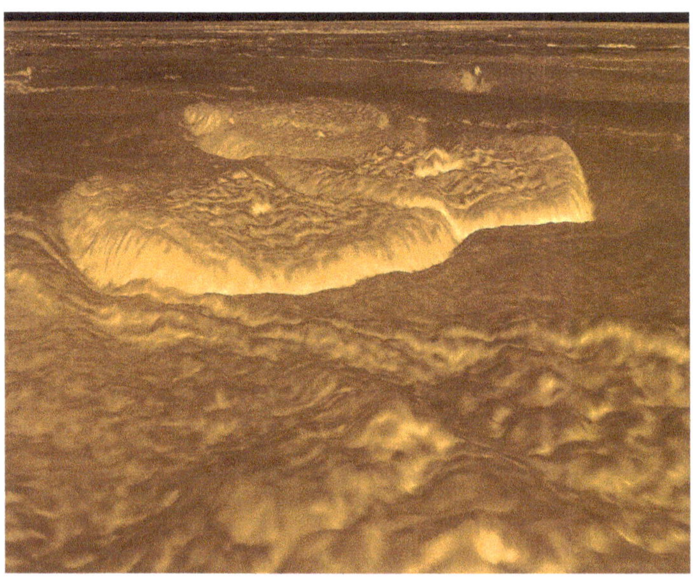

Computer generated perspective view of pancake domes in Venus's Alpha Regio. The domes in this image average 25 km in diameter.

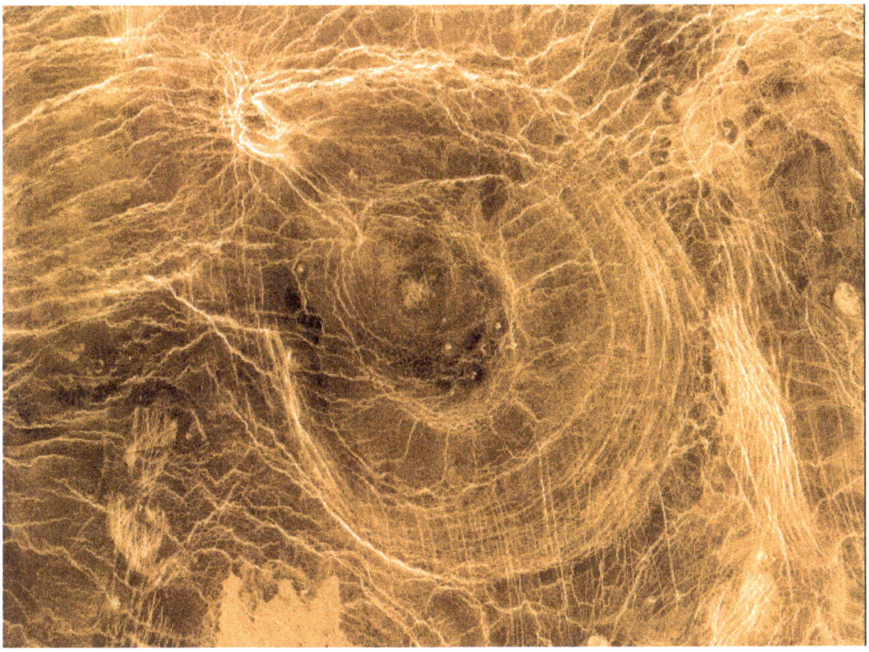

Arachnoid surface feature on Venus.

The surface of Venus is dominated by volcanism. Although Venus is superficially similar to Earth, it seems that the tectonic plates so active in Earth's geology do not exist on Venus. About 80% of the planet consists of a mosaic of volcanic lava plains, dotted with more than a hundred large isolated shield volcanoes, and many hundreds of smaller volcanoes and volcanic constructs such as *coronae*. These are geological features believed to be almost unique to Venus: huge, ring-shaped structures 100–300 kilometers (60–180 mi) across and rising hundreds of meters above the surface. The only other place they have been discovered is on Uranus's moon Miranda. It is believed that they are formed when plumes of rising hot material in the mantle push the crust upwards into a dome shape, which then collapses in the centre as the molten lava cools and leaks out at the sides, leaving a crown-like structure: the corona.

Differences can be seen in volcanic deposits. In many cases, volcanic activity is localized to a fixed source, and deposits are found in the vicinity of this source. This kind of volcanism is called centralized volcanism, in that volcanoes and other geographic features form distinct regions. The second type of volcanic activity is not radial or centralized; flood basalts cover wide expanses of the surface, similar to features such as the Deccan Traps on Earth. These eruptions result in "flow type" volcanoes.

Volcanoes less than 20 kilometres (12 mi) in diameter are very abundant on Venus and they may number hundreds of thousands or even millions. Many appear as flattened domes or ‹pancakes›, thought to be formed in a similar way to shield volcanoes on Earth These pancake dome volcanoes are fairly round features that are less than 1-kilometre (0.62 mi) in height and many times that in width. It is common to find groups of hundreds of these volcanoes in areas called shield fields. The domes of Venus are between 10 and

100 times larger than those formed on Earth. They are usually associated with "coronae" and *tesserae*. The pancakes are thought to be formed by highly viscous, silica-rich lava erupting under Venus's high atmospheric pressure. Domes called scalloped margin domes (commonly called *ticks* because they appear as domes with numerous *legs*), are thought to have undergone mass wasting events such as landslides on their margins. Sometimes deposits of debris can be seen scattered around them.

On Earth, volcanoes are mainly of two types: shield volcanoes and composite or stratovolcanoes. The shield volcanoes, for example those in Hawaii, eject magma from the depths of the Earth in zones called *hot spots*. The lava from these volcanoes is relatively fluid and permits the escape of gases. Composite volcanoes, such as Mount Saint Helens and Mount Pinatubo, are associated with tectonic plates. In this type of volcano, the oceanic crust of one plate slides beneath the other in a subduction zone, together with an inflow of seawater, producing a gummier lava that restricts the exit of the gases, and for that reason, composite volcanoes tend to erupt more violently.

On Venus, where there are no tectonic plates or seawater, volcanoes are of the shield type. Nevertheless, the morphology of the volcanoes of Venus is different. On the Earth, shield volcanoes can be a few tens of kilometers wide and up to 10 kilometers high (6.2 mi) in the case of Mauna Kea, measured from the sea floor. On Venus, these volcanoes can cover hundreds of kilometers in area, but they are relatively flat, with an average height of 1.5 kilometres (0.93 mi).

Other unique features of Venus›s surface are *novae* (radial networks of dikes or grabens) and arachnoids. A nova is formed when large quantities of magma are extruded onto the surface to form radiating ridges and trenches which are highly reflective to radar. These dikes form a symmetrical network around the central point where the lava emerged, where there may also be a depression caused by the collapse of the magma chamber.

Arachnoids are so named because they resemble a spider›s web, featuring several concentric ovals surrounded by a complex network of radial fractures similar to those of a nova. It is not known whether the 250 or so features identified as arachnoids actually share a common origin, or are the result of different geological processes.

TECTONIC ACTIVITY

Despite the fact that Venus appears to have no global plate tectonic system as such, the planet's surface shows various features associated with local tectonic activity. Features such as faults, folds, and volcanoes are present there and may be driven largely by processes in the mantle.

The active volcanism of Venus has generated chains of folded mountains, rift valleys, and terrain known as *tesserae*, a word meaning "floor tiles" in Greek. Tesserae exhibit the effects of eons of compression and tensional deformation.

Unlike those on Earth, the deformations on Venus are directly related to regional dynamic forces within the planet's mantle. Gravitational studies suggest that Venus differs from Earth in lacking an asthenosphere—a layer of lower viscosity and mechanical weakness that allows Earth's crustal tectonic plates to move. The apparent absence of this layer on Venus suggests that the deformation of the Venusian surface must be explained by convective movements within the planet's mantle.

The tectonic deformations on Venus occur on a variety of scales, the smallest of which are related to linear fractures or faults. In many areas these faults appear as networks of parallel lines. Small, discontinuous mountain crests are found which resemble those on the Moon and Mars. The effects of extensive tectonism are shown by the presence of *normal faults*, where the crust has sunk in one area relative to the surrounding rock, and superficial fractures. Radar imaging shows that these types of deformation are concentrated in belts located in the equatorial zones and at high southern latitudes. These belts are hundreds of kilometers wide and appear to interconnect across the whole of the planet, forming a global network associated with the distribution of volcanoes.

The rifts of Venus, formed by the expansion of the lithosphere, are groups of depressions tens to hundreds of meters wide and extending up to 1,000 kilometers in length. The rifts are mostly associated with large volcanic elevations in the form of domes, such as those at Beta Regio, Atla Regio and the western part of Eistla Regio. These highlands seem to be the result of enormous mantle plumes (rising currents of magma) which have caused elevation, fracturing, faulting, and volcanism.

The highest mountain chain on Venus, Maxwell Montes in Ishtar Terra, was formed by processes of compression, expansion, and lateral movement. Another type of geographical feature, found in the lowlands, consists of *ridge belts* elevated several meters above the surface, hundreds of kilometers wide and thousands of kilometers long. Two major concentrations of these belts exist: one in Lavinia Planitia near the southern pole, and the second adjacent to Atalanta Planitia near the northern pole.

Tesserae are found mainly in Aphrodite Terra, Alpha Regio, Tellus Regio and the eastern part of Ishtar Terra (Fortuna Tessera). These regions contain the superimposition and intersection of grabens of different geological units, indicating that these are the oldest parts of the planet. It was once thought that the tesserae were continents associated with tectonic plates like those of the Earth; in reality they are probably the result of floods of basaltic lava forming large plains, which were then subjected to intense tectonic fracturing.

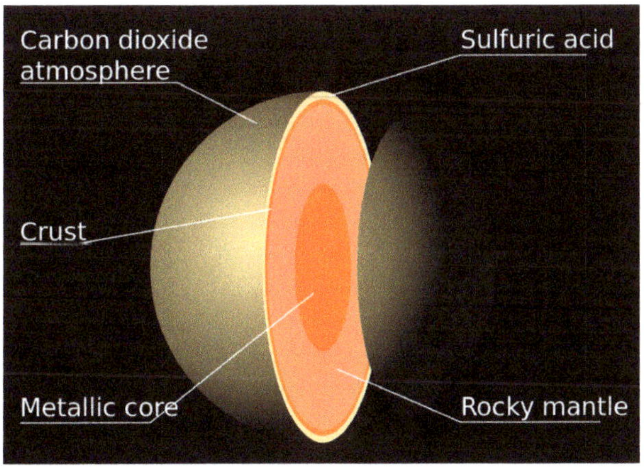

Cutaway diagram of possible internal structure.

MAGNETIC FIELD AND INTERNAL STRUCTURE

Venus's crust appears to be 70 kilometres (43 mi) thick, and composed of silicate rocks. Venus's mantle is approximately 2,840 kilometres (1,760 mi) thick, its chemical composition is probably similar to that of chondrites. Since Venus is a terrestrial planet, it is presumed to have a core made of semisolid iron and nickel with a radius of approximately 3,000 kilometres (1,900 mi).

The unavailability of seismic data from Venus severely limits what can be definitely known about the structure of the planet›s mantle, but models of Earth›s mantle have been modified to make predictions. It›s expected that the uppermost mantle, from about 70 kilometres (43 mi) to 480 kilometres (300 mi) deep is mostly made of the mineral olivine. Descending through the mantle, the chemical composition remains largely the same but at somewhere between about 480 kilometres (300 mi) and 760 kilometres (470 mi), the increasing pressure causes the crystal structure of olivine to change to the more densely packed structure of spinel. Another transition occurs between 760 kilometres (470 mi) and 1,000 kilometres (620 mi) deep, where the material takes on the progressively more compact crystal structures of ilmenite and perovskite, and gradually becomes more like perovskite until the core boundary is reached.

Venus is similar to Earth in size and density, and so probably also in bulk composition, but it does not have a significant magnetic field. Earth's magnetic field is produced by what is known as the core dynamo, consisting of an electrically conducting liquid, the nickel-iron outer core that rotates and is convecting. Venus is expected to have an electrically conductive core of similar composition, and although its rotation period is very long (243.7 Earth days), simulations show that this is adequate to produce a dynamo. This implies that Venus lacks convection in its outer core. Convection occurs when there is a large difference in temperature between the inner and outer part of the core, but since Venus has no plate tectonics to let off heat from the mantle, it is possible that outer core convection is being suppressed by a warm mantle. It›s also possible that Venus may lack a solid inner core for the same reason, if the core is either too hot or is not under enough pressure to allow molten nickel-iron to freeze there.

LAVA FLOWS AND CHANNELS

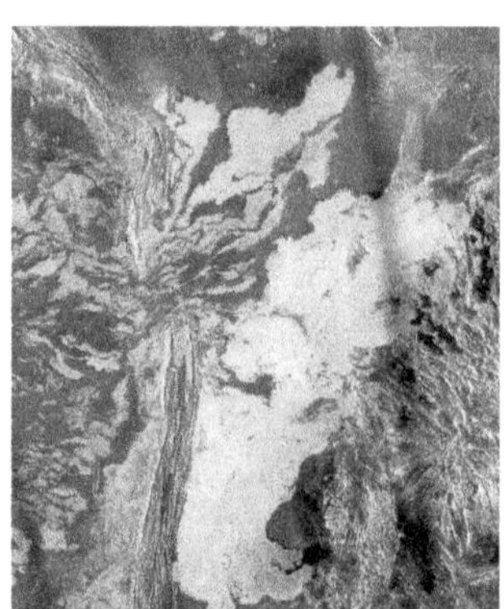

Lava originating from Ammavaru caldera (300 km outside the image) overflowed the ridge left of center and pooled to its right.

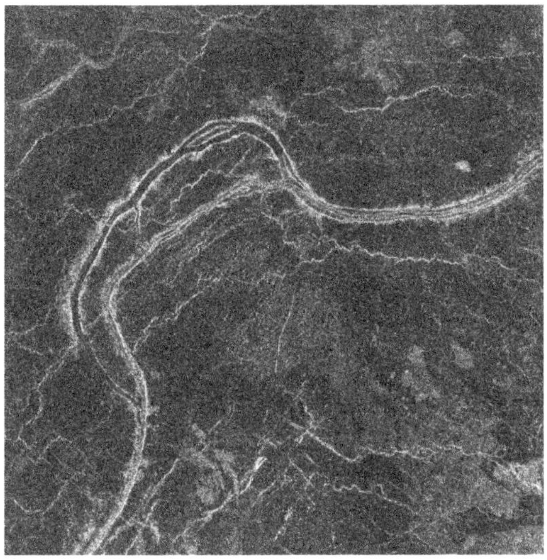

An anastomosing 2-km-wide lava channel in Sedna Planitia.

Lava flows on Venus are often much larger than Earth's, up to several hundred kilometers long and tens of kilometers wide. It is still unknown why these lava fields or *lobate flows* reach such sizes, but it is suggested that they are the result of very large eruptions of basaltic, low-viscosity lava spreading out to form wide, flat plains.

These variations can also reflect differences in lava age and preservation. Channels and lava tubes (channels that have cooled down and over which a dome has formed) are very common on Venus. Two planetary astronomers from the University of Wollongong in Australia, Dr Graeme Melville and Prof. Bill Zealey, researched these lava tubes, using data supplied by NASA, over a number of years and concluded that they were widespread and up to ten times the size of those on the Earth. Melville and Zealey said that the gigantic size of the Venusian lava tubes (tens of meters wide and hundreds of kilometers long) may be explained by the very fluid lava flows together with the high temperatures on Venus, allowing the lava to cool slowly.

For the most part, lava flow fields are associated with volcanoes. The central volcanoes are surrounded by extensive flows that form the core of the volcano. They are also related to fissure craters, coronae, dense clusters of volcanic domes, cones, wells and channels.

Thanks to *Magellan*, more than 200 channels and valley complexes have been identified. The channels were classified as simple, complex, or compound. Simple channels are characterized by a single, long main channel. This category includes rills similar to those found on the Moon, and a new type, called *canali*, consisting of long, distinct channels which maintain their width throughout their entire course. The longest such channel identified (Baltis Vallis) has a length of more than 6,800 kilometres (4,200 mi), about one-sixth of the circumference of the planet.

Complex channels include anastomosed networks, in addition to distribution networks. This type of channel has been observed in association with several impact craters and important lava floods related to major lava flow fields. Compound channels are made of both simple and complex segments. The largest of these channels shows an anastomosed web and modified hills similar to those present on Mars.

Although the shape of these channels is highly suggestive of fluid erosion, there is no evidence that they were formed by water. In fact, there is no evidence of water anywhere on Venus in the last 600 million years. While the most popular theory for the channels› formation is that they are the result of thermal erosion by lava, there are other hypotheses, including that they were formed by heated fluids formed and ejected during impacts.

SURFACE PROCESSES

Water is almost nonexistent on Venus, and thus the only erosive process to be found (apart from thermal erosion by lava flows) is the interaction produced by the atmosphere with the surface. This interaction is present in the ejecta of impact craters expelled onto the surface of Venus. The material ejected during a meteorite impact is lifted to the upper atmosphere, where winds transport the material toward the west. As the material is deposited on the surface, it forms parabola-shaped patterns. This type of deposit can be established on top of various geologic features or lava flows. Therefore, these deposits are the youngest structures on the planet. Images from *Magellan* reveal the existence of more than 60 of these parabola-shaped deposits that are associated with crater impacts.

The ejection material, transported by the wind, is responsible for the process of renovation of the surface at speeds, according to the measurements of the *Venera* soundings, of approximately one metre per second. Given the density of the lower Venusian atmosphere, the winds are more than sufficient to provoke the erosion of the surface and the transportation of fine-grained material. In the regions covered by ejection deposits one may find wind lines, dunes, and yardangs. The wind lines are formed when the wind blows ejection material and volcanic ash, depositing it on top of topographic obstacles such as domes. As a consequence, the leeward sides of domes are exposed to the impact of small grains that remove the surface cap. Such processes expose the material beneath, which has a different roughness, and thus different characteristics under radar, compared to formed sediment.

The dunes are formed by the depositing of particulates that are the size of grains of sand and have wavy shapes. Yardangs are formed when the wind-transported material carves the fragile deposits and produces deep furrows.

The line-shaped patterns of wind associated with impact craters follow a trajectory in the direction of the equator. This tendency suggests the presence of a system of circulation of Hadley cells between medium latitudes and the equator. *Magellan* radar data confirm the existence of strong winds that blow toward the east in the upper surface of Venus, and meridional winds on the surface.

Meteor impacts on Venus have occurred for the last hundreds of millions of years. The superposition of lava flows can be noted. Radar reflection from the oldest lava flows, covered by the newest flows, present distinct intensities. The oldest flows reflect less than the plains that surround the flows. Data from *Magellan* show that the most recent flows are similar to *a a* and *p hoehoe*. However, the oldest lava flows are darker and look like deposits in arid regions of the Earth that have suffered meteor impacts. Chemical and mechanical erosion of the old lava flows is caused by reactions of the surface with the atmosphere in the presence of carbon dioxide and sulfur dioxide (see carbonate-silicate cycle for details). These two gases are the planet's first and third most abundant gases, respectively; the second most abundant gas is inert nitrogen. The reactions probably include the deterioration of silicates by carbon dioxide to produce carbonates and quartz, as well as the deterioration of silicates by sulfur dioxide to produce anhydrate calcium sulfate and carbon dioxide.

ANCIENT LIQUID WATER

NASA's Goddard Institute for Space Studies and others have postulated that Venus may have had a shallow ocean in the past for up to 2 billion years, with as much water as Earth. Depending on the parameters used in their theoretical model, the last liquid water could have evaporated as recently as 715 million years ago. Currently, the only known water on Venus is in the form of a tiny amount of atmospheric vapor (20 ppm). Hydrogen, a component of water, is still being lost to space nowadays as detected by ESA›s Venus Express spacecraft.

REFERENCES

1. de Pater, Imke; Lissauer, Jack J. (2001). Planetary Sciences (First ed.). Cambridge University Press. ISBN 0521482194.
2. Basilevsky, A. T.; J. W. Head III (2003). The surface of Venus (PDF). Reports on Progress in Physics. **66** *(10): 1699–1734. Bibcode:2003RPPh...66.1699B. doi:10.1088/0034-4885/66/10/R04. Archived from the original (PDF) on 2006-03-27.*
3. Ivanov, Mikhail A.; Head, James W. (2010). Scientific Pamphlet, Geologic Map of the Lakshmi Planum Quadrangle (V–7), Venus(PDF). USGS. Retrieved 27 September 2016.
4. Zolotov, M.Y. (1991). Pyrite Stability on the Surface of Venus. Abstracts of the Lunar and Planetary Science Conference. **22**: 1569–1570. Retrieved 27 September 2016.
5. Fegley, Bruce (1997-08-01). Why Pyrite Is Unstable on the Surface of Venus. Icarus. **128** *(2): 474–479. doi:10.1006/icar.1997.5744.*
6. Shepard, Michael K.; Arvidson, Raymond E.; Brackett, Robert A.; Fegley, Bruce (1994-03-15). A ferroelectric model for the low emissivity highlands on Venus. Geophysical Research Letters. **21** *(6): 469–472. doi:10.1029/94GL00392. ISSN 1944-8007.*
7. Treiman, Allan; Harrington, Elise; Sharpton, Virgil (2016-12-01). Venus› radar-bright highlands: Different signatures and materials on Ovda Regio and on Maxwell Montes. Icarus. MicroMars to MegaMars. **280**: 172–182. doi:10.1016/j.icarus.2016.07.001.
8. Wolfram-Alpha: Computational Knowledge Engine. wolframalpha.com.

9. Wolfram-Alpha: Computational Knowledge Engine. wolframalpha.com.
10. Bougher, S. W.; Hunten, D. M.; Philips, R. J.; McKinnon, William B.; Zahnle, Kevin J.; Ivanov, Boris A.; Melosh, H. J. (1997). Venus II – Geology, Geophysics, Atmosphere, and Solar Wind Environment. Tucson: The University of Arizona Press. p. 969. ISBN 0-8165-1830-0.
11. Basilevsky, A. T.; Head, J. W.; Setyaeva, I. V. (1 September 2003). Venus: Estimation of age of impact craters on the basis of degree of preservation of associated radar-dark deposits. Geophys. Res. Lett. **30** *(18): 1950. doi:10.1029/2003GL017504 – via Wiley Online Library.*
12. Kreslavsky, Mikhail A.; Ivanov, Mikhail A.; Head, James W. (21 December 2014). The resurfacing history of Venus: Constraints from buffered crater densities (PDF). Icarus. **250**: 438–450. Retrieved 7 October 2016.
13. Robert G. Strom, Gerald G. Schaber, Douglas D. Dawson The global resurfacing of Venus Journal of Geophysical Research: Planets (1991–2012) Volume 99, Issue E5, pages 10899–10926, 25 May 1994
14. Battaglia, Steven M. (March 2016). Venus: Could Resurfacing Events be Triggered by Sun›s Oscillations through the Galactic Mid-Plane? (PDF). The 47th Lunar and Planetary Science Conference.
15. Stevenson, David J. (15 March 2003). Planetary magnetic fields. Earth and Planetary Science Letters. **208** *(1-2): 1–11. doi:10.1016/S0012-821X(02)01126-3.*
16. Hashimoto, G. L.; Roos-Serote, M.; Sugita, S.; Gilmore, M. S.; Kamp, L. W.; Carlson, R. W.; Baines, K. H. (2008). Felsic highland crust on Venus suggested by Galileo Near-Infrared Mapping Spectrometer data. Journal of Geophysical Research: Planets. **113**: E00B24. Bibcode:2008JGRE..11300B24H. doi:10.1029/2008JE003134.
17. David Shiga (10 October 2007). Did Venus›s ancient oceans incubate life?. New Scientist.
18. Michael J. Way; et al. (26 August 2016). Was Venus the First Habitable World of our Solar System?. Geophysical Research Letters. doi:10.1002/2016GL069790. Retrieved 19 November 2016.
19. Michael Cabbage and Leslie McCarthy (11 August 2016). NASA climate modeling suggests Venus may have been habitable. NASA. Retrieved 19 November 2016.
20. Shannon Hall (10 August 2016). Hellish Venus Might Have Been Habitable for Billions of Years. Scientific American. Retrieved 19 November 2016.
21. Where did Venus›s water go?. European Space Agency. 18 December 2008. Retrieved 19 November 2016.
22. Basilevsky, Alexandr T.; Head, James W. (2003). The surface of Venus. Rep. Prog. Phys. **66** *(10): 1699–1734. Bibcode:2003RPPh...66.1699B. doi:10.1088/0034-4885/66/10/R04.*

23. Bertaux, Jean-Loup; Vandaele, Ann-Carine; Korablev, Oleg; Villard, E.; Fedorova, A.; Fussen, D.; Quémerais, E.; Belyaev, D.; et al. (2007). A warm layer in Venus› cryosphere and high-altitude measurements of HF, HCl, H2O and HDO. Nature. **450** *(7170): 646–649. Bibcode:2007Natur.450..646B. doi:10.1038/nature05974. PMID 18046397.*

CHAPTER 9

Geology of Mars

Chapter Contents

Mars is the fourth planet from the Sun and it orbits the Sun just outside the Earth. Its orbital radius is about 1.5 times that of the Earth-Sun distance. This Earth-Sun distance is generally referred to as 1 astronomical unit (or 1 *AU*). Mars is a smaller planet than the Earth. The size of spherical objects, such as that of the planets, are best described by a mean radius. The mean radius is the average distance from the center to the surface of the planet. Mars' mean radius is about 3,396 *km*, as compared to Earth's, which is 6.378 *km*. Mars' mass is about 6.4×10^{23} *kg*.

INTRODUCTION

The geology of Mars is the scientific study of the surface, crust, and interior of the planet Mars. It emphasizes the composition, structure, history, and physical processes that shape the planet. It is analogous to the field of terrestrial geology. In planetary science, the term *geology* is used in its broadest sense to mean the study of the solid parts of planets and moons. The term incorporates aspects of geophysics, geochemistry, mineralogy, geodesy, and cartography. A neologism, **areology**, from the Greek word *Ar s* (Mars), sometimes appears as a synonym for Mars's geology in the popular media and works of science fiction (e.g. Kim Stanley Robinson's Mars trilogy), but the term is rarely, if ever, used by professional geologists and planetary scientists.

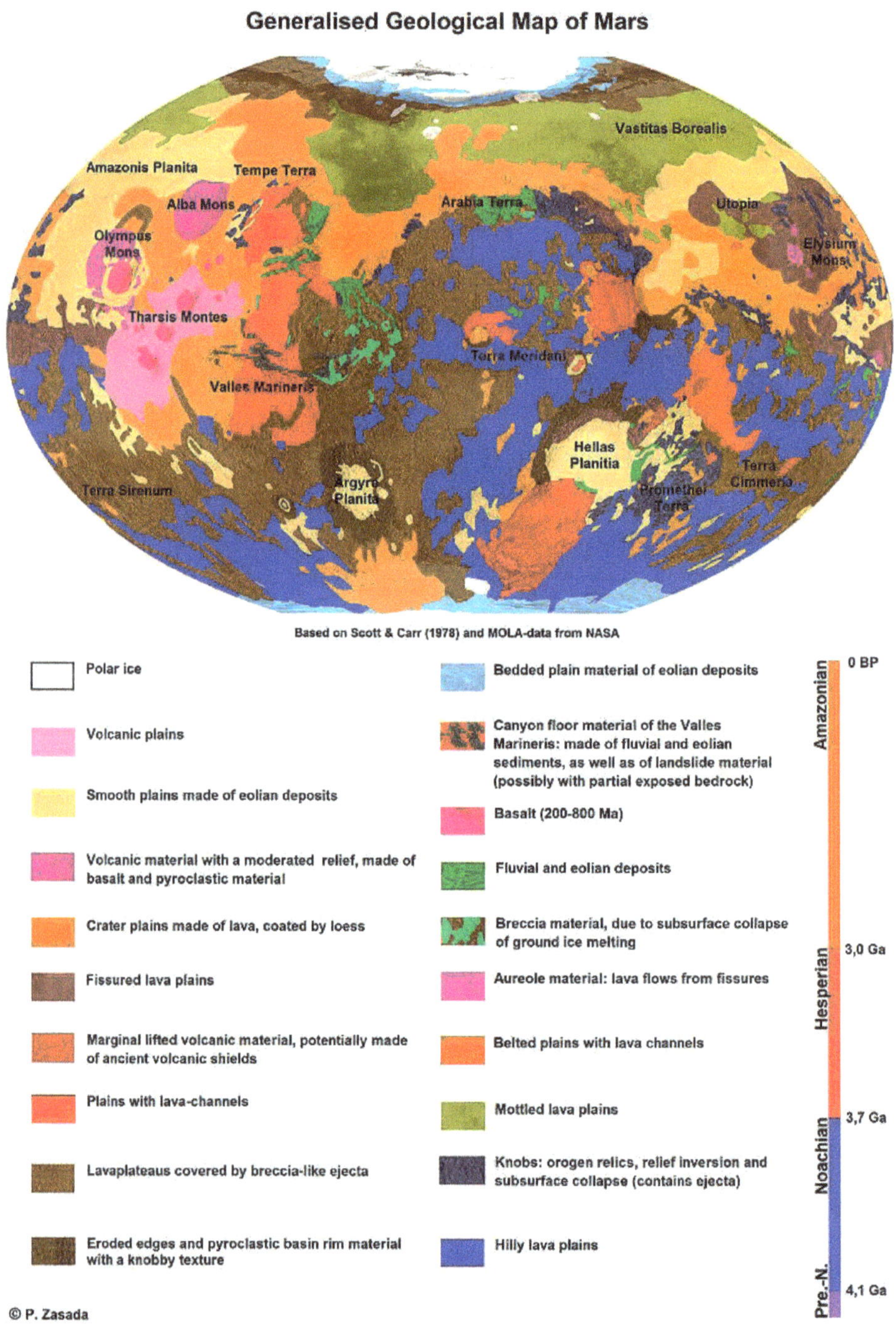

Generalised geological map of Mars

Mars as seen by the Hubble Space Telescope

GEOLOGICAL MAP OF MARS (2014)

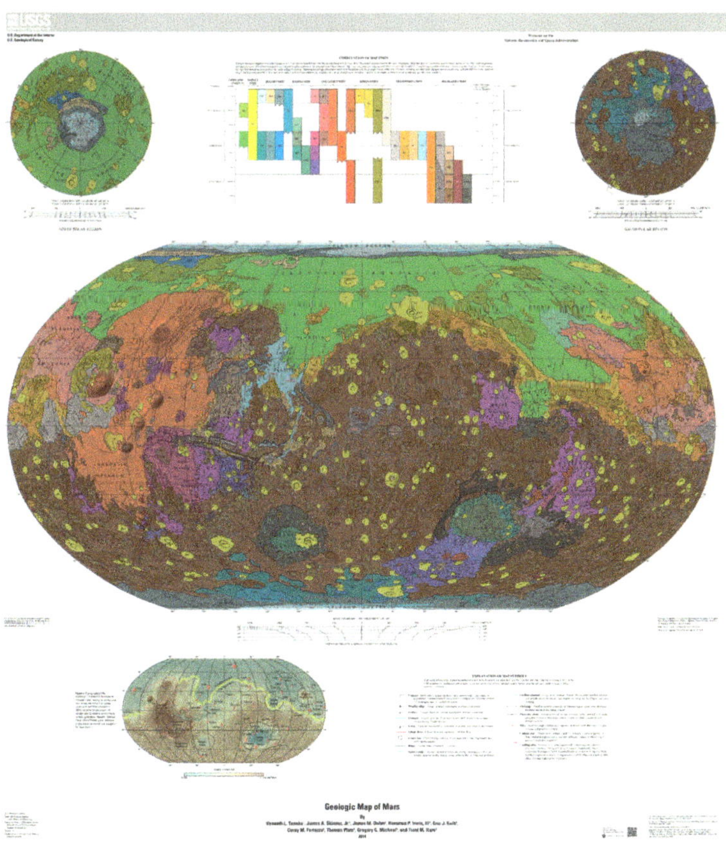

Mars - geologic map (USGS; July 14, 2014)

GLOBAL MARTIAN TOPOGRAPHY AND LARGE-SCALE FEATURES

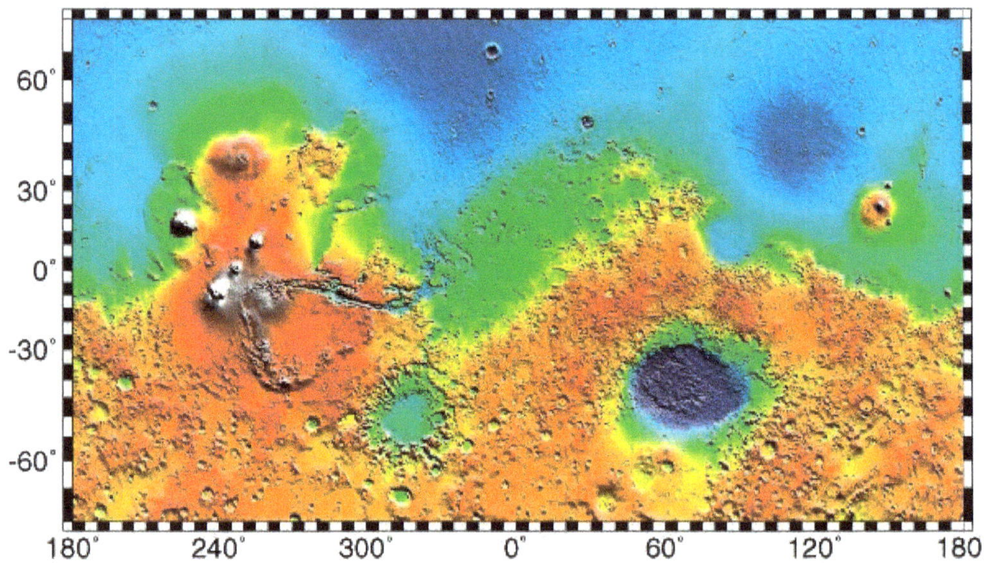

Interactive imagemap of the global topography of Mars, overlain with locations of Mars landers and rovers (Red label = Rover; Blue label = Lander; bold red/blue = currently active). Hover your mouse to see the names of over 25 prominent geographic features, and click to link to them. Coloring of the base map indicates relative elevations, based on data from the Mars Orbiter Laser Altimeter on NASA's Mars Global Surveyor. Reds and pinks are higher elevation (+3 km to +8 km); yellow is 0 km; greens and blues are lower elevation (down to −8 km). Whites (>+12 km) and browns (>+8 km) are the highest elevations. Axes are latitude and longitude; Poles are not shown.

COMPOSITION OF MARS

Mars is a differentiated, terrestrial planet.

Global Physiography

Most of our current knowledge about the geology of Mars comes from studying landforms and relief features (terrain) seen in images taken by orbiting spacecraft. Mars has a number of distinct, large-scale surface features that indicate the types of geological processes that have operated on the planet over time. This section introduces several of the larger physiographic regions of Mars. Together, these regions illustrate how geologic processes involving volcanism, tectonism, water, ice, and impacts have shaped the planet on a global scale.

Hemispheric Dichotomy

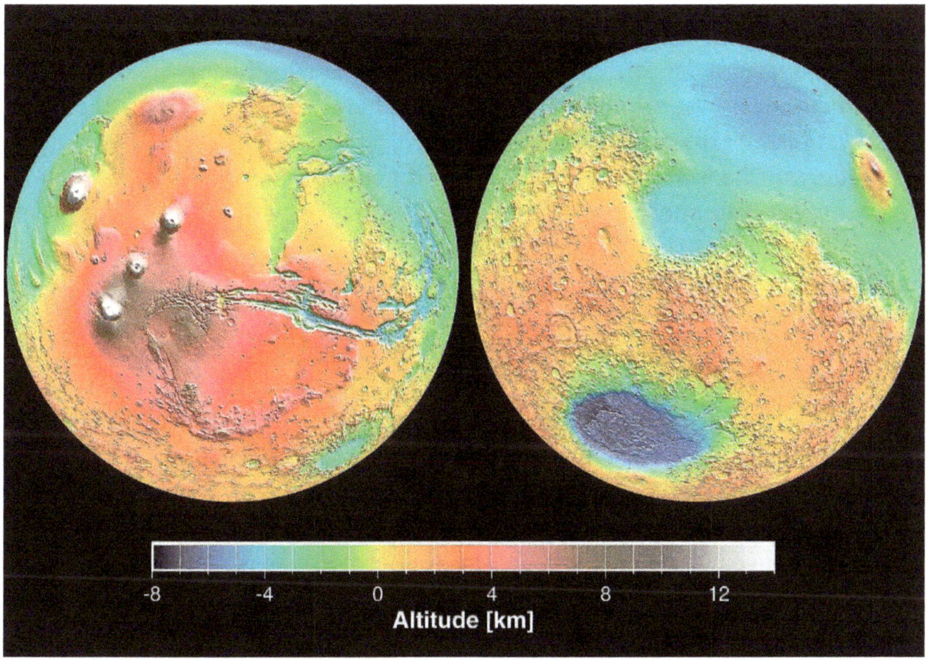

Mars Orbital Laser Altimeter (MOLA) colorized shaded-relief maps showing elevations in the western and eastern hemispheres of Mars. (Left): The western hemisphere is dominated by the Tharsis region (red and brown). Tall volcanoes appear white. Valles Marineris (blue) is the long gash-like feature to the right. (Right): Eastern hemisphere shows the cratered highlands (yellow to red) with the Hellas basin (deep blue/purple) at lower left. The Elysium province is at the upper right edge. Areas north of the dichotomy boundary appear as shades of blue on both maps.

The northern and southern hemispheres of Mars are strikingly different from each other in topography and physiography. This dichotomy is a fundamental global geologic feature of the planet. Simply stated, the northern part of the planet is an enormous topographic depression. About one-third of the planet's surface (mostly in the northern hemisphere) lies 3–6 km lower in elevation than the southern two-thirds. This is a first-order relief feature on par with the elevation difference between Earth's continents and ocean basins. The dichotomy is also expressed in two other ways: as a difference in impact crater density and crustal thickness between the two hemispheres. The hemisphere south of the dichotomy boundary (often called the southern highlands or uplands) is very heavily cratered and ancient, characterized by rugged surfaces that date back to the period of heavy bombardment. In contrast, the lowlands north of the dichotomy boundary have few large craters, are very smooth and flat, and have other features indicating that extensive resurfacing has occurred since the southern highlands formed. The third distinction between the two hemispheres is in crustal thickness. Topographic and geophysical gravity data indicate that the crust in the southern highlands has a maximum thickness of about 58 km (36 mi), whereas crust in the northern lowlands peaks at around 32 km (20 mi) in thickness. The location of the dichotomy boundary varies in latitude across Mars and depends on which of the three physical expressions of the dichotomy is being considered.

The origin and age of the hemispheric dichotomy are still debated. Hypotheses of origin generally fall into two categories: one, the dichotomy was produced by a mega-impact event or several large impacts early in the planet's history (exogenic theories) or two, the dichotomy was produced by crustal thinning in the northern hemisphere by mantle convection, overturning, or other chemical and thermal processes in the planet's interior (endogenic theories). One endogenic model proposes an early episode of plate tectonics producing a thinner crust in the north, similar to what is occurring at spreading plate boundaries on Earth. Whatever its origin, the Martian dichotomy appears to be extremely old. A new theory based on the Southern Polar Giant Impact and validated by the discovery of twelve hemispherical alignments shows that exogenic theories appear to be stronger than endogenic theories and that Mars never had plate tectonics that could modify the dichotomy. Laser altimeter and radar sounding data from orbiting spacecraft have identified a large number of basin-sized structures previously hidden in visual images. Called quasi-circular depressions (QCDs), these features likely represent derelict impact craters from the period of heavy bombardment that are now covered by a veneer of younger deposits. Crater counting studies of QCDs suggest that the underlying surface in the northern hemisphere is at least as old as the oldest exposed crust in the southern highlands. The ancient age of the dichotomy places a significant constraint on theories of its origin.

Tharsis and Elysium Volcanic Provinces

Straddling the dichotomy boundary in Mars's western hemisphere is a massive volcano-tectonic province known as the Tharsis region or the Tharsis bulge. This immense, elevated structure is thousands of kilometers in diameter and covers up to 25% of the planet's surface. Averaging 7–10 km above datum (Martian sea level), Tharsis contains the highest elevations on the planet and the largest known volcanoes in the Solar System. Three enormous volcanoes, Ascraeus Mons, Pavonis Mons, and Arsia Mons (collectively known as the Tharsis Montes), sit aligned NE-SW along the crest of the buldge. The vast Alba Mons (formerly Alba Patera) occupies the northern part of the region. The huge shield volcano Olympus Mons lies off the main buldge, at the western edge of the province. The extreme massiveness of Tharsis has placed tremendous stresses on the planet's lithosphere. As a result, immense extensional fractures (grabens and rift valleys) radiate outward from Tharsis, extending halfway around the planet.

A smaller volcanic center lies several thousand kilometers west of Tharsis in Elysium. The Elysium volcanic complex is about 2,000 kilometers in diameter and consists of three main volcanoes, Elysium Mons, Hecates Tholus, and Albor Tholus. The Elysium group of volcanoes is thought to be somewhat different from the Tharsis Montes, in that development of the former involved both lavas and pyroclastics.

Large Impact Basins

Several enormous, circular impact basins are present on Mars. The largest one that is readily visible is the Hellas basin located in the southern hemisphere. It is the second largest confirmed impact structure on the planet, centered at about 64°E longitude and 40°S latitude. The central part of the basin (Hellas Planitia) is 1,800 km in diameter and surrounded by a broad, heavily eroded annular rim structure characterized by closely spaced rugged irregular mountains (massifs), which probably represent uplifted, jostled blocks of old pre-basin crust. (See Anseris Mons, for example.) Ancient, low-relief volcanic constructs (highland paterae) are located on the northeastern and southwestern portions of the rim. The basin floor contains thick, structurally complex sedimentary deposits that have a long geologic history of deposition, erosion, and internal deformation. The lowest elevations on the planet are located within the Hellas basin, with some areas of the basin floor lying over 8 km below datum.

The two other large impact structures on the planet are the Argyre and Isidis basins. Like Hellas, Argyre (800 km in diameter) is located in the southern highlands and is surrounded by a broad ring of mountains. The mountains in the southern portion of the rim, Charitum Montes, may have been eroded by valley glaciers and ice sheets at some point in Mars›s history. The Isidis basin (roughly 1,000 km in diameter) lies on the dichotomy boundary at about 87°E longitude. The northeastern portion of the basin rim has been eroded and is now buried by northern plains deposits, giving the basin a semicircular outline. The northwestern rim of the basin is characterized by arcuate grabens (Nili Fossae) that are circumferential to the basin. One additional large basin, Utopia, is completely buried by northern plains deposits. Its outline is clearly discernable only from altimetry data. All of the large basins on Mars are extremely old, dating back to the late heavy bombardment. They are thought to be comparable in age to the Imbrium and Orientale basins on the Moon.

Equatorial Canyon System

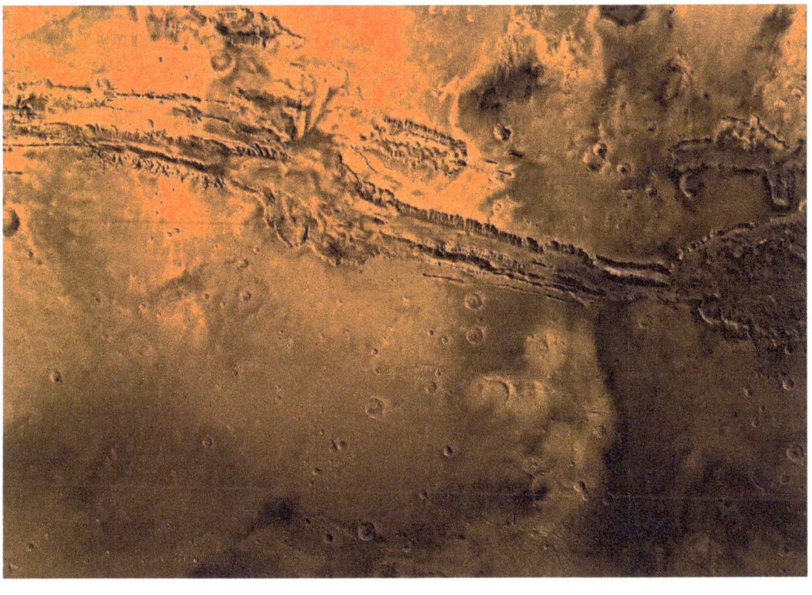

Viking Orbiter 1 view image of Valles Marineris.

Near the equator in the western hemisphere lies an immense system of deep, interconnected canyons and troughs collectively known as the Valles Marineris. The canyon system extends eastward from Tharsis for a length of over 4,000 km, nearly a quarter of the planet's circumference. If placed on Earth, Valles Marineris would span the width of North America. In places, the canyons are up to 300 km wide and 10 km deep. Often compared to Earth's Grand Canyon, the Valles Marineris has a very different origin than its tinier, so-called counterpart on Earth. The Grand Canyon is largely a product of water erosion. The Martian equatorial canyons were of tectonic origin, i.e. they were formed mostly by faulting. They could be similar to the East African Rift valleys. The canyons represent the surface expression of powerful extensional strain in the Martian crust, probably due to loading from the Tharsis bulge.

Chaotic Terrain and Outflow Channels

The terrain at the eastern end of the Valles Marineris grades into dense jumbles of low rounded hills that seem to have formed by the collapse of upland surfaces to form broad, rubble-filled hollows. Called chaotic terrain, these areas mark the heads of huge outflow channels that emerge full size from the chaotic terrain and empty (debouch) northward into Chryse Planitia. The presence of streamlined islands and other geomorphic features indicate that the channels were most likely formed by catastrophic releases of water from aquifers or the melting of subsurface ice. However, these features could also be formed by abundant volcanic lava flows coming from Tharsis. The channels, which include Ares, Shalbatana, Simud, and Tiu Valles, are enormous by terrestrial standards, and the flows that formed them correspondingly immense. For example, the peak discharge required to carve the 28-km-wide Ares Vallis is estimated to have been 500 million cubic feet per second, over ten thousand times the average discharge of the Mississippi River.

Mars Orbital Laser Altimeter (MOLA) derived image of Planum Boreum. Vertical exaggeration is extreme. Note that residual ice cap is only the thin veneer (shown in white) on top of the plateau.

Ice Caps

The polar ice caps are well-known telescopic features of Mars, first identified by Christiaan Huygens in 1672. Since the 1960s, we have known that the seasonal caps (those seen in the telescope to grow and wane seasonally) are composed of carbon dioxide (CO_2) ice that condenses out of the atmosphere as temperatures fall to 148 K, the frost point of CO_2, during the polar wintertime. In the north, the CO_2 ice completely dissipates (sublimes) in summer, leaving behind a residual cap of water (H_2O) ice. At the south pole, a small residual cap of CO_2 ice remains in summer.

Both residual ice caps overlie thick layered deposits of interbedded ice and dust. In the north, the layered deposits form a 3 km-high, 1,000 km-diameter plateau called Planum Boreum. A similar kilometers-thick plateau, Planum Australe, lies in the south. Both plana (the Latin plural of planum) are sometimes treated as synonymous with the polar ice caps, but the permanent ice (seen as the high albedo, white surfaces in images) forms only a relatively thin mantle on top of the layered deposits. The layered deposits probably represent alternating cycles of dust and ice deposition caused by climate changes related to variations in the planet›s orbital parameters over time (see also Milankovitch cycles). The polar layered deposits are some of the youngest geologic units on Mars.

GEOLOGIC HISTORY

Albedo Features

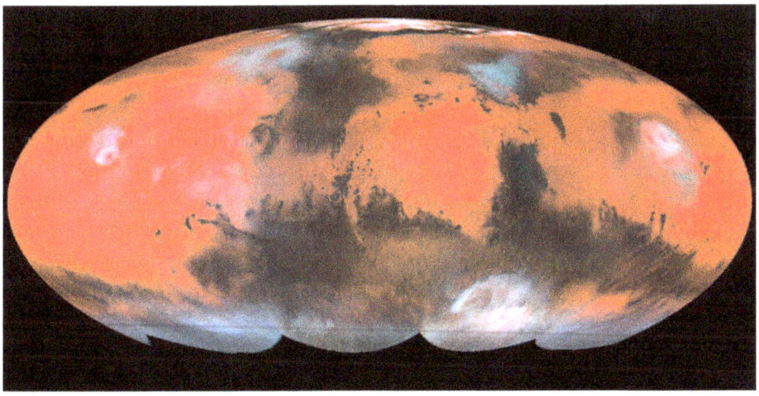

Mollweide projection of albedo features on Mars from Hubble Space Telescope. Bright ochre areas in left, center, and right are Tharsis, Arabia, and Elysium, respectively. The dark region at top center left is Acidalium Planitia. Syrtis Major is the dark area projecting upward in the center right. Note orographic clouds over Olympus and Elysium Montes (left and right, respectively).

No topography is visible on Mars from Earth. The bright areas and dark markings seen through a telescope are albedo features. The bright, red-ochre areas are locations where fine dust covers the surface. Bright areas (excluding the polar caps and clouds) include Hellas, Tharsis, and Arabia Terra. The dark gray markings represent areas that the wind has swept clean of dust, leaving behind the lower layer of dark, rocky material. Dark markings are most distinct in a broad belt from 0° to 40° S latitude. However, the most prominent dark marking, Syrtis Major Planum, is in the northern hemisphere. The classical albedo feature, Mare Acidalium (Acidalia Planitia), is another prominent dark area in the northern hemisphere. A third type of area, intermediate in color and albedo, is also present and thought to represent regions containing a mixture of the material from the bright and dark areas.

IMPACT CRATERS

Impact craters were first identified on Mars by the Mariner 4 spacecraft in 1965. Early observations showed that Martian craters were generally shallower and smoother than lunar craters, indicating that Mars has a more active history of erosion and deposition than the Moon.

In other aspects, Martian craters resemble lunar craters. Both are products of hypervelocity impacts and show a progression of morphology types with increasing size. Martian craters below about 7 km in diameter are called simple craters; they are bowl-shaped with sharp raised rims and have depth/diameter ratios of about 1/5. Martian craters change from simple to more complex types at diameters of roughly 5 to 8 km. Complex craters have central peaks (or peak complexes), relatively flat floors, and terracing or slumping along the inner walls. Complex craters are shallower than simple craters in proportion to their widths, with depth/diameter ratios ranging from 1/5 at the simple-to-complex transition diameter (~7 km) to about 1/30 for a 100-km diameter crater. Another transition occurs at crater diameters of around 130 km as central peaks turn into concentric rings of hills to form multi-ring basins.

Mars has the greatest diversity of impact crater types of any planet in the Solar System. This is partly because the presence of both rocky and volatile-rich layers in the subsurface produces a range of morphologies even among craters within the same size classes. Mars also has an atmosphere that plays a role in ejecta emplacement and subsequent erosion. Moreover, Mars has a rate of volcanic and tectonic activity low enough that ancient, eroded craters are still preserved, yet high enough to have resurfaced large areas of the planet, producing a diverse range of crater populations of widely differing ages. Over 42,000 impact craters greater than 5 km in diameter have been catalogued on Mars, and the number of smaller craters is probably innumerable. The density of craters on Mars is highest in the southern hemisphere, south of the dichotomy boundary. This is where most of the large craters and basins are located.

Crater morphology provides information about the physical structure and composition of the surface and subsurface at the time of impact. For example, the size of central peaks in Martian craters is larger than comparable craters on Mercury or the Moon. In addition, the central peaks of many large craters on Mars have pit craters at their summits. Central pit craters are rare on the Moon but are very common on Mars and the icy satellites of the outer Solar System. Large central peaks and the abundance of pit craters probably indicate the presence of near-surface ice at the time of impact. Polewards of 30 degrees of latitude, the form of older impact craters is rounded out (softened) by acceleration of soil creep by ground ice. The most notable difference between Martian craters and other craters in the Solar System is the presence of lobate (fludized) ejecta blankets. Many craters at equatorial and mid-latitudes on Mars have this form of ejecta morphology, which is thought to arise when the impacting object melts ice in the subsurface. Liquid water in the ejected material forms a muddy slurry that flows along the surface, producing the characteristic lobe shapes. The crater Yuty is a good example of a rampart crater, which is so called because of the rampart-like edge to its ejecta blanket.

HiRISE image of simple rayed crater on southeastern flank of Elysium Mons.

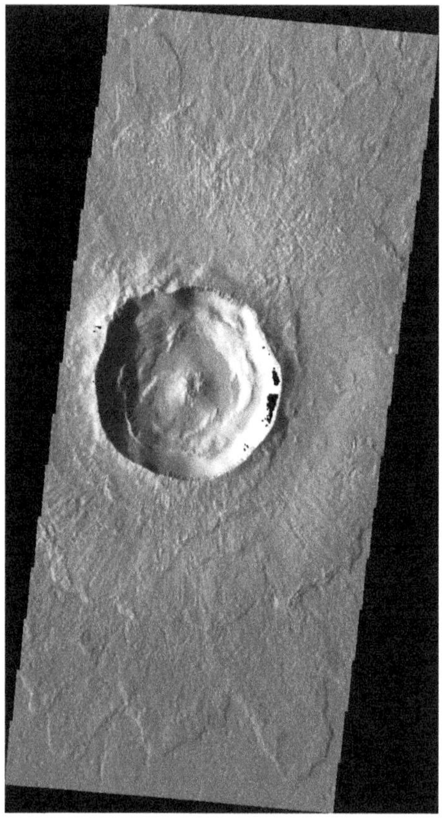

THEMIS image of complex crater with fluidized ejecta. Note central peak with pit crater.

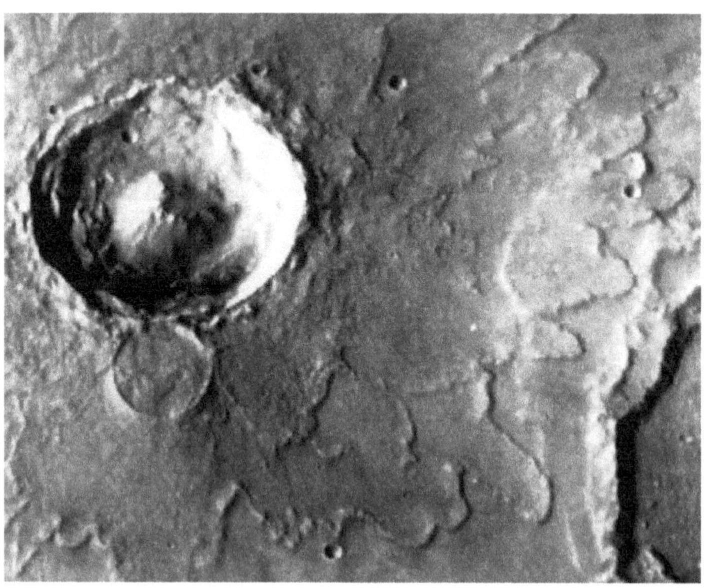

Viking orbiter image of Yuty crater showing lobate ejecta.

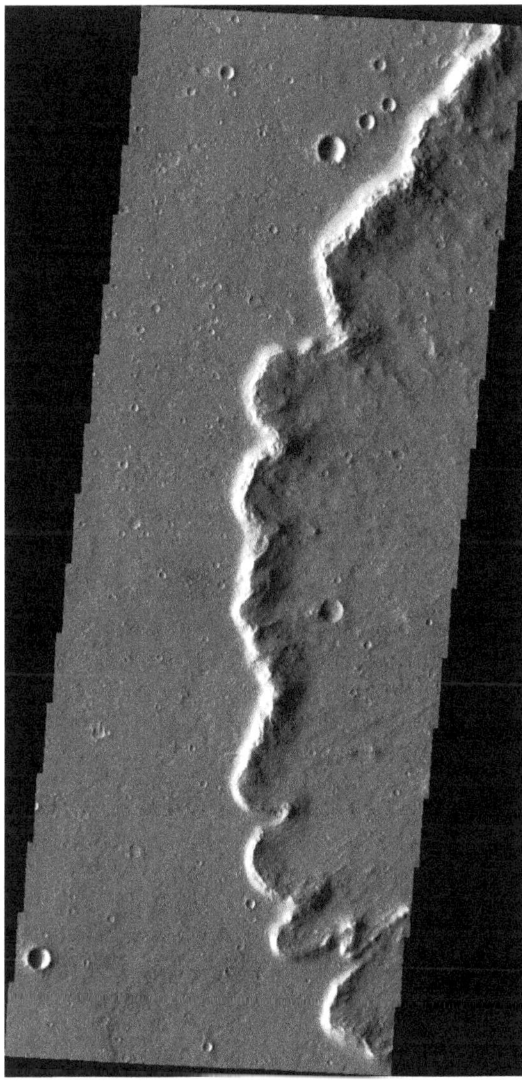

THEMIS close-up view of ejecta from 17-km diameter crater at 21°S, 285°E. Note prominent rampart.

Martian craters are commonly classified by their ejecta. Craters with one ejecta layer are called single-layer ejecta (SLE) craters. Craters with two superposed ejecta blankets are called double-layer ejecta (DLE) craters, and craters with more than two ejecta layers are called multiple-layered ejecta (MLE) craters. These morphological differences are thought to reflect compositional differences (i.e. interlayered ice, rock, or water) in the subsurface at the time of impact.

Pedestal crater in Amazonis quadrangle as seen by HiRISE.

Martian craters show a large diversity of preservational states, from extremely fresh to old and eroded. Degraded and infilled impact craters record variations in volcanic, fluvial, and eolian activity over geologic time. Pedestal craters are craters with their ejecta sitting above the surrounding terrain to form raised platforms. They occur because the crater's ejecta forms a resistant layer so that the area nearest the crater erodes more slowly than the rest of the region. Some pedestals are hundreds of meters above the surrounding area, meaning that hundreds of meters of material were eroded away. Pedestal craters were first observed during the Mariner 9 mission in 1972.

VOLCANISM

First X-ray diffraction view of Martian soil - CheMin analysis reveals feldspar, pyroxenes, olivine and more (Curiosity rover at "Rocknest, October 17, 2012).

Volcanic structures and landforms cover large portions of the Martian surface. The most conspicuous volcanoes on Mars occur in Tharsis and Elysium. Geologists think one of the reasons that volcanoes on Mars are able to grow so large is because Mars has sparsely few tectonic boundaries compared to Earth. Lava from a stationary hot spot is able to accumulate at one location on the surface for many hundreds of millions of years.

On October 17, 2012, the Curiosity rover on the planet Mars at "Rocknest performed the first X-ray diffraction analysis of Martian soil. The results from the rover›s CheMin analyzer revealed the presence of several minerals, including feldspar, pyroxenes and olivine, and suggested that the Martian soil in the sample was similar to the weathered basaltic soils of Hawaiian volcanoes.In July 2015, the same rover identified tridymite in a rock sample from Gale Crater, leading scientists to believe that silicic volcanism might have played a much more prevalent role in the planet's volcanic history than previously thought.

SEDIMENTOLOGY

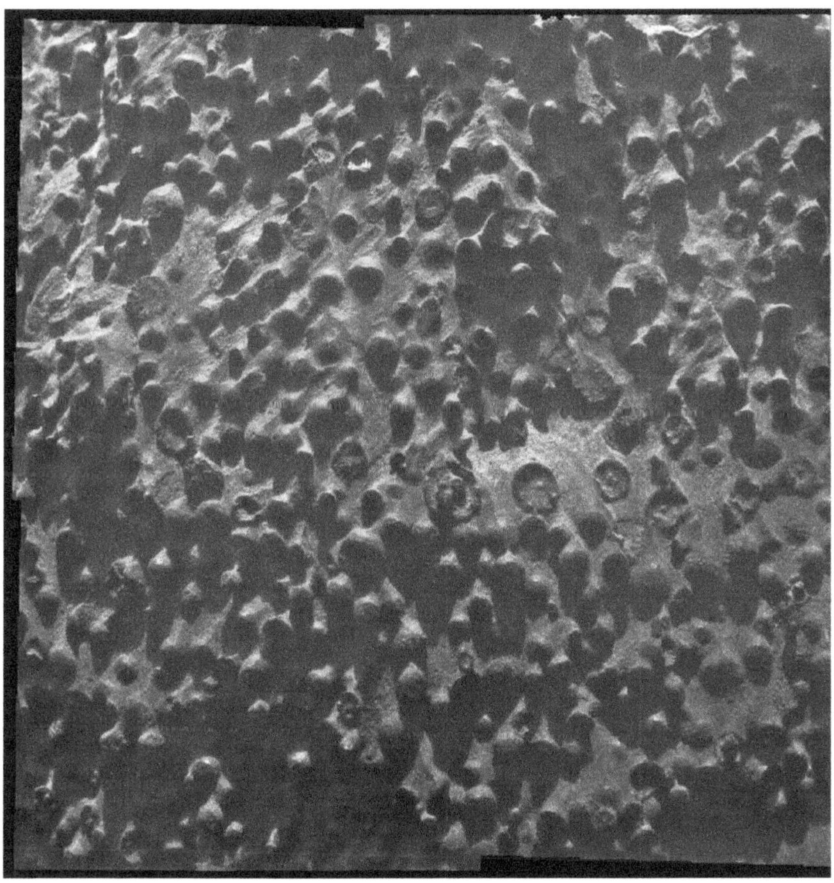

Collection of spheres, each about 3 mm in diameter as seen by *Opportunity* rover.

Flowing water appears to have been common on the surface of Mars at various points in its history, and especially on ancient Mars. Many of these flows carved the surface, forming valley networks and producing sediment. This sediment has been redeposited in a wide variety of wet environments, including in alluvial fans, meandering channels, deltas, lakes, and perhaps even oceans. The processes of deposition and transportation are associated with gravity. Due to gravity, related differences in water fluxes and flow speeds, inferred from grain size distributions, Martian landscapes were created by different environmental conditions. Nevertheless, there are other ways of estimating the amount of water on ancient Mars (see: Water on Mars). Groundwater has been implicated in the cementation of aeolian sediments and the formation and transport of a wide variety of sedimentary minerals including clays, sulphates and hematite.

When the surface has been dry, wind has been a major geomorphic agent. Wind driven sand bodies like megaripples and dunes are extremely common on the modern Martian surface, and Opportunity has documented abundant aeolian sandstones on its traverse. Ventifacts, like Jake Matijevic (rock), are an other aeolian landform on the Martian Surface.

A wide variety of other sedimentological facies are also present locally on Mars, including glacial deposits, hot springs, dry mass movement deposits (especially landslides), and cryogenic and periglacial material, amongst many others.

COMMON SURFACE FEATURES

Groundwater on Mars

One group of researchers proposed that some of the layers on Mars were caused by groundwater rising to the surface in many places, especially inside of craters. According to the theory, groundwater with dissolved minerals came to the surface, in and later around craters, and helped to form layers by adding minerals (especially sulfate) and cementing sediments. This hypothesis is supported by a groundwater model and by sulfates discovered in a wide area. At first, by examining surface materials with Opportunity Rover, scientists discovered that groundwater had repeatedly risen and deposited sulfates. Later studies with instruments on board the Mars Reconnaissance Orbiter showed that the same kinds of materials exist in a large area that included Arabia.

INTERESTING GEOLOGICAL FEATURES

Avalanches

On February 19, 2008, images obtained by the HiRISE camera on the Mars Reconnaissance Orbiter showed a spectacular avalanche, in which debris thought to be fine-grained ice, dust, and large blocks fell from a 700-metre (2,300 ft) high cliff. Evidence of the avalanche included dust clouds rising from the cliff afterwards. Such geological events are theorized to be the cause of geologic patterns known as slope streaks.

Image of the February 19, 2008 Mars avalanche captured by the Mars Reconnaissance Orbiter.

Closer shot of the avalanche.

Dust clouds rise above the 700-metre (2,300 ft) deep cliff.

A photo with scale demonstrates the size of the avalanche.

Possible Caves

NASA scientists studying pictures from the Odyssey spacecraft have spotted what might be seven caves on the flanks of the Arsia Mons volcano on Mars. The pit entrances measure from 100 to 252 metres (328 to 827 ft) wide and they are believed to be at least 73 to 96 metres (240 to 315 ft) deep. See image below: the pits have been informally named (A) Dena, (B) Chloe, (C) Wendy, (D) Annie, (E) Abby (left) and Nikki, and (F) Jeanne. Because light did not reach the floor of most of the pits, it is likely that they extend much deeper than these lower estimates. Dena's floor was observed and found to be 130m deep. Further investigation suggested that these were not necessarily lava tube "skylights". Review of the images has resulted in yet more discoveries of deep pits.

A cave on Mars (“Jeanne”) as seen by the Mars Reconnaissance Orbiter.

HiRISE closeup of Jeanne showing afternoon illumination of the east wall of the shaft.

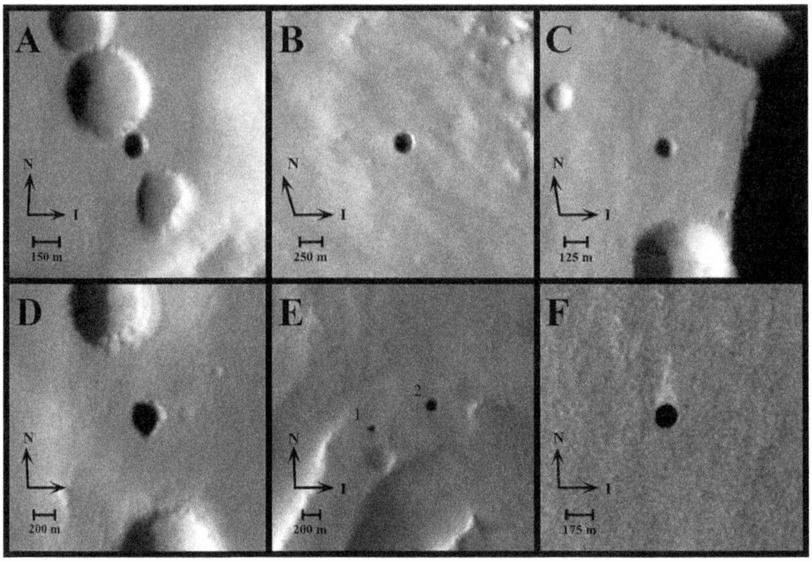

THEMIS image of cave entrances on Mars.

It has been suggested that human explorers on Mars could use lava tubes as shelters. The caves may be the only natural structures offering protection from the micrometeoroids, UV radiation, solar flares, and high energy particles that bombard the planet's surface.

Inverted Relief

Some areas of Mars show inverted relief, where features that were once depressions, like streams, are now above the surface. It is believed that materials like large rocks were deposited in low-lying areas. Later, wind erosion removed much of the surface layers, but left behind the more resistant deposits. Other ways of making inverted relief might be lava flowing down a stream bed or materials being cemented by minerals dissolved in water. On Earth, materials cemented by silica are highly resistant to all kinds of erosional forces. Examples of inverted channels on Earth are found in the Cedar Mountain Formation near Green River, Utah. Inverted relief in the shape of streams are further evidence of water flowing on the Martian surface in past times. Inverted relief in the form of stream channels suggest that the climate was different—much wetter—when the inverted channels were formed.

In an article published in January 2010, a large group of scientists endorsed the idea of searching for life in Miyamoto Crater because of inverted stream channels and minerals that indicated the past presence of water.

Images of other examples of inverted terrain are shown below from various parts of Mars.

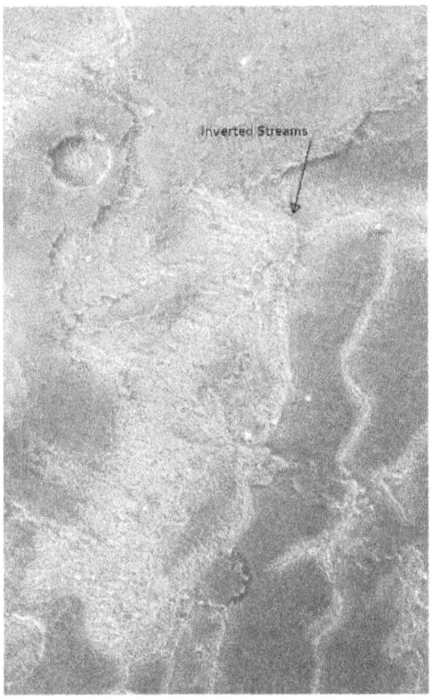

Inverted Streams near Juventae Chasma, as seen by Mars Global Surveyor. These streams begin at the top of a ridge then run together.

Inverted Channel with many branches in Syrtis Major quadrangle.

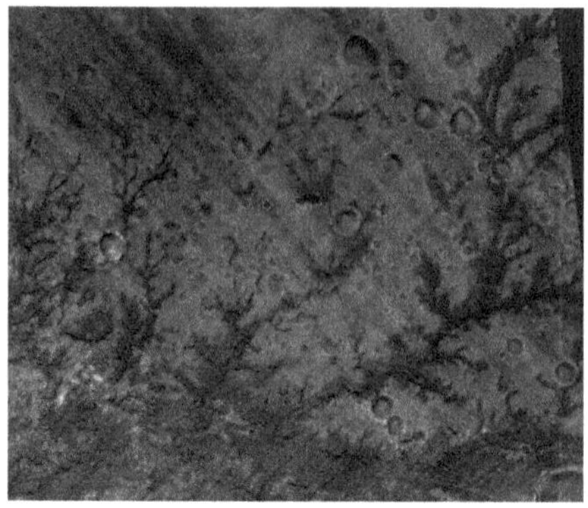

Inverted Stream Channels in Antoniadi Crater, as seen by HiRISE. Image in Syrtis Major quadrangle.

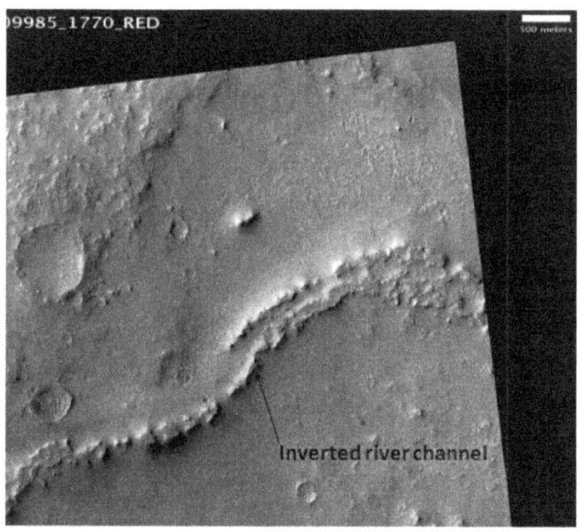

Inverted Channel in Miyamoto Crater, as seen by HiRISE. Image is located in Margaritifer Sinus quadrangle. The scale bar is 500 meters long.

REFERENCES

1. P. Zasada (2013) Generalised Geological Map of Mars, 1:140.000.000, Source Link.
2. Greeley, Ronald (1993). Planetary landscapes (2nd ed.). New York: Chapman & Hall. p. 1. ISBN 0-412-05181-8.
3. Quinion, M. (1996). World Wide Words Website, http://www.worldwidewords.org/turnsofphrase/tp-are1.htm.
4. Carr, M.H., USGS, Personal Communication, September 13, 2010.
5. Tanaka, Kenneth L.; Skinner, James A., Jr.; Dohm, James M.; Irwin, Rossman

P., III; Kolb, Eric J.; Fortezzo, Corey M.; Platz, Thomas; Michael, Gregory G.; Hare, Trent M. (July 14, 2014). Geologic Map of Mars - 2014. USGS. Retrieved July 22, 2014.

6. Krisch, Joshua A. (July 22, 2014). Brand New Look at the Face of Mars. New York Times. Retrieved July 22, 2014.

7. Staff (July 14, 2014). Mars - Geologic map - Video (00:56). USGS. Retrieved July 22, 2014.

8. Watters, Thomas R.; McGovern, Patrick J.; Irwin Iii, Rossman P. (2007). Hemispheres Apart: The Crustal Dichotomy on Mars (PDF). Annu. Rev. Earth Planet. Sci. **35** *(1): 621–652 [624, 626]. Bibcode*:2007AREPS..35..621W. doi:10.1146/annurev.earth.35.031306.140220.

9. Zuber, M. T.; Solomon, SC; Phillips, RJ; Smith, DE; Tyler, GL; Aharonson, O; Balmino, G; Banerdt, WB; et al. (2000). Internal Structure and Early Thermal Evolution of Mars from Mars Global Surveyor Topography and Gravity. Science. **287** *(5459): 1788–93. Bibcode*:2000Sci...287.1788Z. doi:10.1126/science.287.5459.1788. PMID 10710301.

10. Neumann, G. A. (2004). Crustal structure of Mars from gravity and topography. Journal of Geophysical Research. **109** *(E8). Bibcode*:2004JGRE..10908002N. doi:10.1029/2004JE002262.

11. Wilhelms, D.E.; Squyres, S.W. (1984). The Martian Hemispheric Dichotomy May Be Due to a Giant Impact. Nature. **309** *(5964): 138–140. Bibcode*:1984Natur.309..138W. doi:10.1038/309138a0.

12. Frey, Herbert; Schultz, Richard A. (1988). Large impact basins and the mega-impact origin for the crustal dichotomy on Mars". Geophysical Research Letters. **15** *(3): 229–232. Bibcode*:1988GeoRL..15..229F. doi:10.1029/GL015i003p00229.

13. Andrews-Hanna, J.C.; et al. (2008). The Borealis Basin and the Origin of the Martian Crustal Dichotomy. Nature. **453** *(7199). pp. 1212–5; see p. 1212. Bibcode*:2008Natur.453.1212A. doi:10.1038/nature07011. PMID 18580944.

14. Wise, Donald U.; Golombek, Matthew P.; McGill, George E. (1979). Tectonic Evolution of Mars. Journal of Geophysical Research. **84** *(B14): 7934–7939. Bibcode*:1979JGR....84.7934W. doi:10.1029/JB084iB14p07934.

15. Elkins-Tanton, Linda T.; Hess, Paul C.; Parmentier, E. M. (2005). Possible formation of ancient crust on Mars through magma ocean processes (PDF). Journal of Geophysical Research. **110** *(E12): E120S01. Bibcode*:2005JGRE..11012S01E. doi:10.1029/2005JE002480.

16. Sleep, Norman H. (1994). Martian plate tectonics. Journal of Geophysical Research. **99** *(E3): 5639–5655. Bibcode*:1994JGR....99.5639S. doi:10.1029/94JE00216.

17. Leone, Giovanni; Tackley, Paul J.; Gerya, Taras V.; May, Dave A.; Zhu, Guizhi (2014-12-28). Three-dimensional simulations of the southern polar giant impact hypothesis for the origin of the Martian dichotomy. Geophysical Research Letters. **41** *(24): 2014GL062261.* doi:10.1002/2014GL062261. ISSN 1944-8007.

18. Leone, Giovanni (2016-01-01). Alignments of volcanic features in the southern

hemisphere of Mars produced by migrating mantle plumes. Journal of Volcanology and Geothermal Research. **309**: 78–95. doi:10.1016/j.jvolgeores.2015.10.028.

19. O'Rourke, Joseph G.; Korenaga, Jun (2012-11-01). Terrestrial planet evolution in the stagnant-lid regime: Size effects and the formation of self-destabilizing crust. Icarus. **221** *(2): 1043–1060. doi:10.1016/j.icarus.2012.10.015.*
20. Wong, Teresa; Solomatov, Viatcheslav S (2015-07-02). Towards scaling laws for subduction initiation on terrestrial planets: constraints from two-dimensional steady-state convection simulations. Progress in Earth and Planetary Science. **2** *(1).* doi:10.1186/s40645-015-0041-x. ISSN 2197-4284.
21. Watters, T.R.; McGovern, Patrick J.; Irwin, R.P. (2007). Hemispheres Apart: The Crustal Dichotomy on Mars. Annu. Rev. Earth Planet. Sci. **35**: 630–635. Bibcode:2007AREPS..35..621W. doi:10.1146/annurev.earth.35.031306.140220.
22. Solomon, S. C.; Aharonson, O; Aurnou, JM; Banerdt, WB; Carr, MH; Dombard, AJ; Frey, HV; Golombek, MP; et al. (2005). New Perspectives on Ancient Mars. Science. **307** *(5713): 1214–20. Bibcode*:2005Sci...307.1214S. doi:10.1126/science.1101812. PMID 15731435.
23. Solomon, Sean C.; Head, James W. (1982). Evolution of the Tharsis Province of Mars: The Importance of Heterogeneous Lithospheric Thickness and Volcanic Construction. J. Geophys. Res. **87** *(B12): 9755–9774. Bibcode*:1982JGR....87.9755S. doi:10.1029/JB087iB12p09755.
24. Carr, M.H (2007). Mars: Surface and Interior in Encyclopedia of the Solar System, 2nd ed., McFadden, L.-A. et al. Eds. Elsevier: San Diego, CA, p.319
25. Cattermole, Peter John (2001). Mars: the mystery unfolds. Oxford: Oxford University Press. p. 71. ISBN 0-19-521726-8.
26. Boyce, J.M. (2008) The Smithsonian Book of Mars; Konecky&Konecky: Old Saybrook, CT, p. 13.
27. Carr, M.H.; Saunders, R.S.; Strom R.G. (1984). Geology of the Terrestrial Planets; NASA Scientific and Technical Information Branch: Washington DC, 1984, p. 223. http://www.lpi.usra.edu/publications/books/geologyTerraPlanets/
28. Hartmann 2003, pp. 70–73
29. Kargel, J.S.; Strom, R.G. (1992). Ancient Glaciation on Mars. Geology. **20** *(1): 3–7. Bibcode*:1992Geo....20....3K. doi:10.1130/0091-7613(1992)020<0003:AGOM>2.3.CO;2.
30. Kargel, J.S. (2004) Mars: A Warmer Wetter Planet; Springer-Praxis: London, p. 52.
31. Leone, Giovanni (2014-05-01). A network of lava tubes as the origin of Labyrinthus Noctis and Valles Marineris on Mars. Journal of Volcanology and Geothermal Research. **277**: 1–8. doi:10.1016/j.jvolgeores.2014.01.011.
32. Baker, Victor R. (2001). Water and the Martian Landscape (PDF). Nature. **412** *(6843). pp. 228–36; see p. 231 Fig. 5.* doi:10.1038/35084172. PMID 11449284.
33. Sheehan, W. (1996). The Planet Mars: A History of Observation & Discovery; University of Arizona Press: Tucson, p. 25. http://www.uapress.arizona.edu/

onlinebks/mars/contents.htm.

34. Leighton, R.B.; Murray, B.C. (1966). Behavior of Carbon Dioxide and Other Volatiles on Mars. Science. **153** *(3732): 136–144. Bibcode*:1966Sci...153..136L. doi:10.1126/science.153.3732.136. PMID 17831495.

35. Arvidson, Raymond E.; Guinness, Edward A.; Dale-Bannister, Mary A.; Adams, John; Smith, Milton; Christensen, Philip R.; Singer, Robert B. (1989). Nature and Distribution of Surficial Deposits in Chryse Planitia and Vicinity, Mars. J. Geophys. Res. **94** *(B2): 1573–1587. Bibcode*:1989JGR....94.1573A. doi:10.1029/JB094iB02p01573.

36. Leighton, R.B.; Murray, B.C.; Sharp, R.P.; Allen, J.D.; Sloan, R.K. (1965). Mariner IV Photography of Mars: Initial Results. Science. **149** *(3684): 627–630. Bibcode*:1965Sci...149..627L. doi:10.1126/science.149.3684.627. PMID 17747569.

37. Leighton, R.B.; Horowitz, NH; Murray, BC; Sharp, RP; Herriman, AH; Young, AT; Smith, BA; Davies, ME; Leovy, CB (1969). Mariner 6 and 7 Television Pictures: Preliminary Analysis. Science. **166** *(3901): 49–67. Bibcode*:1969Sci...166...49L. doi:10.1126/science.166.3901.49. PMID 17769751.

38. Pike, R.J. (1980). Formation of Complex Impact Craters: Evidence from Mars and Other Planets. Icarus. **43** *(1): 1*–19 . Bibcode:1980Icar...43....1P. doi:10.1016/0019-1035(80)90083-4.

39. Carr 2006, pp. 24–27

40. Strom, R.G.; Croft, S.K.; Barlow, N.G. (1992). The Martian Impact Cratering Record. In Kieffer, H.H.; Jakosky, B.M.; Snyder, C.W.; et al. Mars. Tucson: University of Arizona Press. pp. 384–385. ISBN 978-0-8165-1257-7.

41. Barlow, N.G. (1988). Crater Size-Frequency Distributions and a Revised Martian Relative Chronology. Icarus. **75** *(2): 285–305. Bibcode*:1988Icar...75..285B. doi:10.1016/0019-1035(88)90006-1.

42. Hale, W.S.; Head, J.W. (1981). Lunar Planet. Sci. XII, pp. 386-388. (abstract 1135). http://www.lpi.usra.edu/meetings/lpsc1981/pdf/1135.pdf

43. Squyres, Steven W., and Michael H. Carr. Geomorphic evidence for the distribution of ground ice on Mars. Science 231.4735 (1986): 249-252.

44. Walter S. Kiefer (2004). Maximum Impact - Impact Craters in the Solar System. NASA Solar System Exploration. Retrieved 2007-05-14.

45. Hartmann 2003, pp. 99–100

46. Viking Orbiter Views Of Mars. NASA. Retrieved 2007-03-16.

47. Boyce, J.M. The Smithsonian Book of Mars; Konecky&Konecky: Old Saybrook, CT, 2008, p. 203.

48. Barlow, N.G.; Boyce, Joseph M.; Costard, Francois M.; Craddock, Robert A.; Garvin, James B.; Sakimoto, Susan E. H.; Kuzmin, Ruslan O.; Roddy, David J.; Soderblom, Laurence A. (2000). Standardizing the Nomenclature of Martian Impact Crater Ejecta Morphologies. J. Geophys. Res. **105** *(E11): 26,733–8. Bibcode*:2000JGR...10526733B. doi:10.1029/2000JE001258.

49. Nadine Barlow. Stones, Wind and Ice. Lunar and Planetary Institute. Retrieved

2007-03-15.

50. http://hirise.lpl.eduPSP_008508_1870
51. Bleacher, J. and S. Sakimoto. Pedestal Craters, A Tool For Interpreting Geological Histories and Estimating Erosion Rates. LPSC
52. Pedestal Craters in Utopia | Mars Odyssey Mission THEMIS
53. Brown, Dwayne (October 30, 2012). NASA Rover›s First Soil Studies Help Fingerprint Martian Minerals. NASA. Retrieved October 31, 2012.
54. Wolpert, Stuart (August 9, 2012). UCLA scientist discovers plate tectonics on Mars. Yin, An. UCLA. Retrieved August 11, 2012.
55. NASA News (22 June 2016), NASA Scientists Discover Unexpected Mineral on Mars, NASA Media, retrieved 23 June 2016
56. Craddock, R.A.; Howard, A.D. (2002). The case for rainfall on a warm, wet early Mars. J. Geophys. Res. **107** *(E11). Bibcode*:2002JGRE..107.5111C. doi:10.1029/2001je001505.
57. Carr, M. 2006. The Surface of Mars. Cambridge University Press. ISBN 978-0-521-87201-0
58. Grotzinger, J. and R. Milliken (eds.) 2012. Sedimentary Geology of Mars. SEPM
59. Salese, F., G. Di Achille, A. Neesemann, G. G. Ori, and E. Hauber (2016), Hydrological and sedimentary analyses of well-preserved paleofluvial-paleolacustrine systems at Moa Valles, Mars, J. Geophys. Res. Planets, 121, 194–232, doi:10.1002/2015JE004891.
60. Patrick Zasada (2013/14): Gradation of extraterrestrial fluvial sediments – related to the gravity. - Z. geol. Wiss. **41/42** (3): 167-183. Abstract
61. Opportunity Rover Finds Strong Evidence Meridiani Planum Was Wet. Retrieved July 8, 2006.
62. S. W. Squyres and A. H. Knoll, Sedimentary Geology at Meridiani Planum, Mars, Elsevier, Amsterdam, ISBN 978-0-444-52250-4 (2005); reprinted from Earth and Planetary Science Letters, Vol. 240, No. 1 (2005).
63. Zasada, P., 2013: Entstehung des Marsgesteins Jake Matijevic. – Sternzeit, issue 2/2013: 98 ff. (in German language).
64. Andrews-Hanna, J. C.; Phillips, R. J.; Zuber, M. T. (2007). Meridiani Planum and the global hydrology of Mars. Nature. **446** *(7132): 163–166. Bibcode*:2007Natur.446..163A. doi:10.1038/nature05594. PMID 17344848.
65. Andrews; Hanna, J. C.; Zuber, M. T.; Arvidson, R. E.; Wiseman, S. M. (2010). Early Mars hydrology: Meridiani playa deposits and the sedimentary record of Arabia Terra. J. Geophys. Res. **115**: E06002. Bibcode:2010JGRE..115.6002A. doi:10.1029/2009JE003485.
66. Grotzinger, J. P.; et al. (2005). Stratigraphy and sedimentology of a dry to wet eolian depositional system, Burns formation, Meridiani Planum, Mars. Earth Planet. Sci. Lett. **240**: 11–72. Bibcode:2005E&PSL.240...11G. doi:10.1016/j.epsl.2005.09.039.

67. McLennan, S. M.; et al. (2005). Provenance and diagenesis of the evaporitebearing Burns formation, Meridiani Planum, Mars. Earth Planet. Sci. Lett. **240**: 95–121. Bibcode:2005E&PSL.240...95M. doi:10.1016/j.epsl.2005.09.041.

68. Squyres, S. W.; Knoll, A. H. (2005). Sedimentary rocks at Meridiani Planum: Origin, diagenesis, and implications for life on Mars. Earth Planet. Sci. Lett. **240**: 1–10. Bibcode:2005E&PSL.240....1S. doi:10.1016/j.epsl.2005.09.038.

69. Squyres, S. W.; et al. (2006). Two years at Meridiani Planum: Results from the Opportunity rover. Science. **313**: 1403–1407. doi:10.1126/science.

70. M. Wiseman, J. C. Andrews-Hanna, R. E. Arvidson3, J. F. Mustard, K. J. Zabrusky DISTRIBUTION OF HYDRATED SULFATES ACROSS ARABIA TERRA USING CRISM DATA: IMPLICATIONS FOR MARTIAN HYDROLOGY. 42nd Lunar and Planetary Science Conference (2011) 2133.pdf

71. DiscoveryChannel.ca - Mars avalanche caught on camera

72. Rincon, Paul (March 17, 2007). ›Cave entrances› spotted on Mars. BBC News.

73. Shiga, David (August 2007). Strange Martian feature not a ‹bottomless› cave after all. New Scientist. Retrieved 2010-07-01.

74. Teen project one-ups NASA, finds hole in Mars cave. AFP. 2010-06-23. Retrieved 2010-07-01.

75. Thompson, Andrea (2009-10-26). Mars Caves Might Protect Microbes (or Astronauts). Space.com. Retrieved 2010-07-01.

76. HiRISE | Inverted Channels North of Juventae Chasma (PSP_006770_1760). Hirise.lpl.arizona.edu. Retrieved 2012-01-16.

77. Newsom, Horton E.; Lanza, Nina L.; Ollila, Ann M.; Wiseman, Sandra M.; Roush, Ted L.; Marzo, Giuseppe A.; Tornabene, Livio L.; Okubo, Chris H.; et al. (2010). Inverted channel deposits on the floor of Miyamoto crater, Mars. Icarus. **205** *(1): 64–72. Bibcode:*2010Icar..205...64N. doi:10.1016/j.icarus.2009.03.030.

Index

P

S

T

V

W